The National Science Foundation's Materials Research Science and Engineering Centers Program

LOOKING BACK, MOVING FORWARD

MRSEC Impact Assessment Committee
Solid State Sciences Committee
Board on Physics and Astronomy
Division on Engineering and Physical Sciences

NATIONAL RESEARCH COUNCIL
OF THE NATIONAL ACADEMIES

THE NATIONAL ACADEMIES PRESS
Washington, D.C.
www.nap.edu

THE NATIONAL ACADEMIES PRESS 500 Fifth Street, N.W. Washington, DC 20001

NOTICE: The project that is the subject of this report was approved by the Governing Board of the National Research Council, whose members are drawn from the councils of the National Academy of Sciences, the National Academy of Engineering, and the Institute of Medicine. The members of the committee responsible for the report were chosen for their special competences and with regard for appropriate balance.

This study was supported by Grant No. DMR-0446470 between the National Academy of Sciences and the National Science Foundation. Any opinions, findings, conclusions, or recommendations expressed in this publication are those of the author(s) and do not necessarily reflect the views of the organizations or agencies that provided support for the project.

International Standard Book Number-13: 978-0-309-10961-1
International Standard Book Number-10: 0-309-10961-2

Cover: An electric field can cause a polymer film to be unstable. If, in addition, the polymer does not stick to the surface of the underlying support, the polymer will withdraw from the surface, like water on a nonstick pan. These two factors produce an instability that can be seen with an optical microscope by the interference colors. The instability, in this case, has caused the formation of this unusual structure, since the electric field that was used was not uniform across the film. The finger-like texture can be used to measure the properties of the polymer. Courtesy of T. Xu, University of Massachusetts at Amherst.

Copies of this report are available from the National Academies Press, 500 Fifth Street, N.W., Lockbox 285, Washington, DC 20055; (800) 624-6242 or (202) 334-3313 (in the Washington metropolitan area), http://www.nap.edu; and the Board on Physics and Astronomy, National Research Council, 500 Fifth Street, N.W., Washington, DC 20001, http://www.national-academies.org/bpa.

Printed in the United States of America

THE NATIONAL ACADEMIES

Advisers to the Nation on Science, Engineering, and Medicine

The **National Academy of Sciences** is a private, nonprofit, self-perpetuating society of distinguished scholars engaged in scientific and engineering research, dedicated to the furtherance of science and technology and to their use for the general welfare. Upon the authority of the charter granted to it by the Congress in 1863, the Academy has a mandate that requires it to advise the federal government on scientific and technical matters. Dr. Ralph J. Cicerone is president of the National Academy of Sciences.

The **National Academy of Engineering** was established in 1964, under the charter of the National Academy of Sciences, as a parallel organization of outstanding engineers. It is autonomous in its administration and in the selection of its members, sharing with the National Academy of Sciences the responsibility for advising the federal government. The National Academy of Engineering also sponsors engineering programs aimed at meeting national needs, encourages education and research, and recognizes the superior achievements of engineers. Dr. Charles M. Vest is president of the National Academy of Engineering.

The **Institute of Medicine** was established in 1970 by the National Academy of Sciences to secure the services of eminent members of appropriate professions in the examination of policy matters pertaining to the health of the public. The Institute acts under the responsibility given to the National Academy of Sciences by its congressional charter to be an adviser to the federal government and, upon its own initiative, to identify issues of medical care, research, and education. Dr. Harvey V. Fineberg is president of the Institute of Medicine.

The **National Research Council** was organized by the National Academy of Sciences in 1916 to associate the broad community of science and technology with the Academy's purposes of furthering knowledge and advising the federal government. Functioning in accordance with general policies determined by the Academy, the Council has become the principal operating agency of both the National Academy of Sciences and the National Academy of Engineering in providing services to the government, the public, and the scientific and engineering communities. The Council is administered jointly by both Academies and the Institute of Medicine. Dr. Ralph J. Cicerone and Dr. Charles M. Vest are chair and vice chair, respectively, of the National Research Council.

www.national-academies.org

MRSEC IMPACT ASSESSMENT COMMITTEE

MATTHEW V. TIRRELL, University of California at Santa Barbara, *Chair*
KRISTI S. ANSETH, University of Colorado at Boulder
MEIGAN ARONSON, Brookhaven National Laboratory
DAVID M. CEPERLEY, University of Illinois at Urbana-Champaign
PAUL M. CHAIKIN, New York University
RONALD C. DAVIDSON, Princeton University
DUANE B. DIMOS, Sandia National Laboratories
FRANCIS J. DiSALVO, Cornell University
EDITH M. FLANIGEN, UOP, Inc. (retired)
THOMAS F. KUECH, University of Wisconsin-Madison
DIANDRA L. LESLIE-PELECKY, University of Nebraska-Lincoln
BRUCE H. MARGON, University of California at Santa Cruz
ANDREW MILLIS, Columbia University
VENKATESH NARAYANAMURTI, Harvard University
RALPH G. NUZZO, University of Illinois at Urbana-Champaign
DOUGLAS D. OSHEROFF, Stanford University
STUART PARKIN, IBM Almaden Research Center
JULIA M. PHILLIPS, Sandia National Laboratories
LYLE H. SCHWARTZ, Air Force Office of Scientific Research (retired)
ELI YABLONOVITCH, University of California at Los Angeles

NEIL E. PATON, LiquidMetal Technologies, Consultant

Staff

DONALD C. SHAPERO, Director, Board on Physics and Astronomy
TIMOTHY I. MEYER, Senior Program Officer
DAVID B. LANG, Research Associate
VAN AN, Financial Associate

SOLID STATE SCIENCES COMMITTEE

PETER F. GREEN, University of Michigan, *Chair*
BARBARA JONES, IBM Almaden Research Center, *Vice Chair*
COLLIN L. BROHOLM, Johns Hopkins University
ELBIO DAGOTTO, Oak Ridge National Laboratory and University of Tennessee
DUANE B. DIMOS, Sandia National Laboratories
JAMES P. EISENSTEIN, California Institute of Technology
SHARON C. GLOTZER, University of Michigan
MARC A. KASTNER, Massachusetts Institute of Technology
STEVEN A. KIVELSON, University of California at Los Angeles
SIDNEY R. NAGEL, University of Chicago
MONICA OLVERA DE LA CRUZ, Northwestern University
ARTHUR P. RAMIREZ, Lucent Technologies, Inc.
A. DOUGLAS STONE, Yale University
ANTOINETTE "TONI" TAYLOR, Los Alamos National Laboratory

Staff

DONALD C. SHAPERO, Director, Board on Physics and Astronomy
NATALIA J. MELCER, Program Officer
CARYN J. KNUTSEN, Senior Program Assistant

Preface

The MRSEC Impact Assessment Committee was convened by the National Research Council in response to an informal request from the National Science Foundation. Charged to examine the impact of the Materials Research Science and Engineering Centers program (MRSEC program) and to provide guidance for the future (see Appendix A), the committee included experts from across materials research as well as several from outside the field (see Appendix G for biographical sketches of the committee members).

The committee describes its analysis in this report at three different levels of detail in order to make the analysis accessible to the broadest possible audience. The Executive Summary provides a brief summary of the report. The Overview describes the complete chain of reasoning and includes all of the findings, conclusions, and recommendations. Chapters 1 through 6 then present detailed discussions and evidence.

In preparing its report, the committee found it necessary to distinguish among three types of key statements. All appear in boldface within this report but are to be distinguished as follows:

- *General finding:* A nontrivial observation that, in the committee's judgment, arises from the evidence examined in the course of its work. These general findings express general principles that are not unique to the MRSEC program performance and impact assessment.
- *Conclusion:* A nontrivial observation that the committee derived during

its work that pertains directly to the MRSEC program's performance and impact assessment.

- *Recommendation:* An action item assigned to specific entities that the committee believes will enhance the future performance and impact of the MRSEC program for materials research.

The committee thanks its generous hosts at each of its site visits (Boston University, California Institute of Technology, Harvard University, Massachusetts Institute of Technology, Michigan State University, University of California at San Diego, University of California at Santa Barbara, University of Florida, University of Michigan, University of Southern California, and University of Southern Mississippi); these half-day meetings were an invaluable data-gathering tool for the committee. The warm hospitality provided an environment for frank discussion and insightful suggestions that contributed to the committee's understanding of the issues. At each of its meetings, many invited experts gave testimony on their experiences working in materials research (see Appendix B). The committee greatly appreciates the time and effort that these individuals put into preparing their remarks.

The committee gratefully acknowledges the thoughtful and very helpful participation of the staff from the National Research Council's Board on Science Education, including Jean Moon, Andrew Shouse, and Yan Liu. These experts helped the committee to collect and analyze data on the education and outreach activities at Materials Research Science and Engineering Centers as well as understand the frontiers of research in science education.

Matthew V. Tirrell, *Chair*
MRSEC Impact Assessment Committee

Acknowledgment of Reviewers

This report has been reviewed in draft form by individuals chosen for their diverse perspectives and technical expertise, in accordance with procedures approved by the National Research Council's Report Review Committee. The purpose of this independent review is to provide candid and critical comments that will assist the institution in making its published report as sound as possible and to ensure that the report meets institutional standards for objectivity, evidence, and responsiveness to the study charge. The review comments and draft manuscript remain confidential to protect the integrity of the deliberative process. We wish to thank the following individuals for their review of this report:

Elihu Abrahams, Rutgers, The State University of New Jersey,
Paul A. Fleury, Yale University,
Jerry Gollub, Haverford College,
Fiona Goodchild, California NanoSystems Institute,
Arthur F. Hebard, University of Florida,
Joseph S. Hezir, EOP Group, Inc.,
Marc A. Kastner, Massachusetts Institute of Technology,
Linda (Lee) J. Magid, University of Tennessee,
Christopher Monroe, University of Michigan,
Donald W. Murphy, Bell Laboratories, Lucent Technologies,
N.P. Ong, Princeton University,
Thomas Russell, University of Massachusetts at Amherst,

Dan J. Thoma, Los Alamos National Laboratory, and
Julia R. Weertman, Northwestern University.

Although the reviewers listed above have provided many constructive comments and suggestions, they were not asked to endorse the conclusions or recommendations, nor did they see the final draft of the report before its release. The review of this report was overseen by Donald M. Tennant, Cornell University. Appointed by the National Research Council, he was responsible for making certain that an independent examination of this report was carried out in accordance with institutional procedures and that all review comments were carefully considered. Responsibility for the final content of this report rests entirely with the authoring committee and the institution.

Contents

Executive Summary

The purpose of this study is to:

1. Assess the performance and impact of the National Science Foundation's Materials Research Science and Engineering Centers program (MRSEC program); and
2. On the basis of current trends and needs in materials and condensed matter research, recommend future directions and roles for the program.

To address this task, the National Research Council's MRSEC Impact Assessment Committee—comprising representatives of universities both with and without Materials Research Science and Engineering Centers (MRSECs), industry, and national laboratories—employed four in-person meetings, four whole-committee teleconferences, extensive questionnaires to and telephone interviews with National Science Foundation (NSF) and university personnel, and visits to current, former, and would-be MRSEC sites. Four working subcommittees, which often met independently, addressed issues associated with research, education and outreach, industrial outreach, and facilities and management. This Executive Summary presents the full committee's principal findings and recommendations.

The nature of materials research demands mechanisms to support interdisciplinary collaboration for the conception and execution of ideas and for developing the capabilities to sustain our nation's competitiveness in the production of new technology and products based on advances in materials science and engineering. This work often is conducted over a very long timescale, and new materials tend to have far-reaching implications for many other fields, from medicine to high-energy physics to the economy. The task at hand was to assess the relative performance and impact of MRSEC-supported activities in comparison with other mechanisms for support and to recommend a robust strategy for the future of the program.

MRSECs have enormous perceived impact.

Conclusion: MRSEC awards continue to be in great demand. The intense competition for them within the community indicates a strong perceived value. These motivations include:

- **The ability to pursue interdisciplinary, collaborative research;**
- **The resources to provide an interdisciplinary training experience for the future scientific and technical workforce from undergraduate to postdoctoral researchers;**
- **Block funding at levels that enable more rapid response to new ideas, and that support higher-risk projects, than is possible with single-investigator grants;**
- **The leverage and motivation MRSECs provide in producing increased institutional, local, and/or state support for materials research;**
- **The perceived distinction that the presence of a MRSEC gives to the materials research enterprise of an institution, thus attracting more quality students and junior faculty; and**
- **The infrastructure that MRSECs can provide to organize and manage facilities and educational and industrial outreach.**

The committee pursued several comprehensive exercises to measure the impact of MRSECs. Constructing algorithms to distinguish the MRSEC-enabled results from others was complicated by the following features:

- MRSEC participants are supported by many funding sources;
- MRSEC participants engage in multiple activities with multiple collaborators;
- Average performance often does not capture the full impact of a portfolio of efforts; and
- MRSECs are intended to enhance the conditions for conceiving of research and education activities, and yet most measures of impact examine the results from the execution of these activities.

Conclusion: The committee examined the performance and impact of MRSEC activities over the past decade in the areas of research, facilities, education and outreach, and industrial collaboration and technology transfer. The MRSEC program has had important impacts of the same high standard

of quality as those of other multi-investigator or individual-investigator programs. Although the committee was largely unable to attribute observed impacts uniquely to the MRSEC program, MRSECs generally mobilize efforts that would not have occurred otherwise.

MRSECs conduct and publish research with characteristics similar to those of other programs. The shared-facilities element of MRSECs represents a significant portion of the NSF investment in midsize facilities for materials research; moreover, the MRSEC program offers one of the few mechanisms for investment in operations and maintenance of shared facilities. The MRSEC education and outreach programs clearly benefit from the sharing and pooling of resources; improvements by NSF and the participating communities are needed, however. Although industrial collaborations that take place within the MRSEC framework are of a similar character as those elsewhere, the activities initiated by MRSECs generally represent efforts that would not have occurred otherwise.

Conclusion: The effectiveness of MRSECs has been reduced in recent years as a result of increasing requirements without a commensurate increase in resources. Increasing the mean grant size is necessary to allow the program to fulfill its important mission goals.

Average funding for these centers, in inflation-adjusted dollars, has declined in the past decade by up to 10 percent. A key element of the MRSEC program is the participation of graduate student researchers. When the program budget history is compared with the increasing costs of graduate education, the trends are even more dramatic. The decline in funding has been particularly detrimental to efforts to build and maintain the advanced instrumentation necessary for leading-edge materials research. Another decade of similar decreases will undermine the ability of the MRSEC program to make future valuable contributions. In addition, the program's responsibilities for industrial partnership and for education and outreach have increased, as has the number of MRSECs, whereas the MRSEC program itself has remained at a relatively constant budget level.

Recommendation: To respond to changes in the budgetary landscape and changes in the nature of materials research in the coming decade, NSF should restructure the MRSEC program to allow more efficient use and leveraging of resources. The new program should fully invest in centers of excellence as well as in stand-alone teams of researchers.

Resources for basic research, especially in materials research, have not kept pace with overall economic growth in the past decade. Expectations for the range and extent of impacts enabled by NSF's programs have also changed. And materials research has continued to mature as a discipline. The MRSEC program can be positioned to better facilitate advances in research in the next decade by focusing its resources on targeted, specific objectives and by increasing flexibility to allow specialization based on the strengths at individual centers. The committee developed one detailed vision of an approach for achieving these objectives.

Two funding mechanisms could be created, under the auspices of the NSF Division of Materials Research: one (Materials Centers of Excellence, or MCEs) would support several coordinated teams of interdisciplinary research groups, carry out educational and industrial outreach, and support state-of-the-art facilities. The second element (Materials Research Groups, or MRGs) would support interdisciplinary research groups that do not have separately mandated educational and industrial activities or facilities. The committee envisioned a revenue-neutral transition to its formulation of the program, although this restructuring would allow NSF to focus more resources on the program in the future.

The key element of this proposal is its holistic approach to a restructuring of the MRSEC program in order to balance the concentration of resources optimally on key topics while preserving breadth in the overall portfolio. The rationale for this shift is to centralize the value-added activities at appropriately funded centers without losing the benefits of interdisciplinary research being done by smaller groups of researchers. To do so, smaller groups (MRGs) would be formed with more flexibility and without some of the responsibilities of the MCEs; conversely, the responsibilities for educational and industrial outreach and facilities development would be taken up by the MCEs as part of their missions. MCEs should not, however, be viewed as more permanent institutions than the current MRSECs, and, in particular, NSF should create a review mechanism for evaluating the research of the research groups within MCEs on some common, comparative, competitive basis with the research outputs of the MRGs. The MCEs would shoulder more of the educational and industrial outreach and facilities development and maintenance responsibilities on behalf of the entire materials research community. Employing a common process and criteria for the review of research while restructuring to distribute responsibilities more effectively will keep the overall portfolio vibrant, competitive, and better matched with the objectives and current budget of the MRSEC program.

Conclusion: NSF encourages MRSECs to operate as a national network. Although some efforts have been made in that direction, the committee did not observe strong cooperation among the discrete centers of the program. The MRSEC program is thus missing a clear opportunity to leverage resources and thereby strengthen the materials research enterprise as a whole.

The opportunity to leverage the combined resources of the MRSEC program is significant. The centers could expedite the pace of the overall research effort by taking advantage of tools and talents distributed throughout the program. Such initiatives, however, are best launched from the centers and the researchers themselves.

Building the integrated capabilities of materials research centers into a cooperating network would strengthen materials science and engineering in the United States as a discipline and as a factor in U.S. competitiveness.

Overview

This Overview reviews the key elements of the report, summarizes each chapter, and details the committee's full set of findings and recommendations.

BACKGROUND

The National Science Foundation's (NSF's) Materials Research Science and Engineering Centers (MRSECs) trace their origin to the Interdisciplinary Laboratories (IDLs) created by the Advanced Research Projects Agency in the 1960s. Initiated in 1994, the MRSECs represent the latest in a series of centers designed to foster organized group research on materials in the academic community. After more than a decade, it is appropriate to examine the MRSEC program in present and future contexts. The National Research Council was asked to carry out such an examination and to:

1. Assess the performance and impact of the National Science Foundation's Materials Research Science and Engineering Centers program (MRSEC program); and
2. On the basis of current trends and needs in materials and condensed matter research, recommend future directions and roles for the program.

The MRSEC Impact Assessment Committee, with representatives of universities (both with and without MRSECs), industry, and national laboratories, employed four in-person meetings, four whole-committee teleconferences, extensive questionnaires to and telephone interviews with NSF and university personnel, and visits to current, former, and would-be MRSEC sites. Four working subcommittees, which often met independently, addressed issues of research, education and outreach, industrial outreach, and facilities and management. This Overview presents the outcome of this study and serves as a map to the more detailed exposition that follows in the body of the report.

The MRSEC technical agenda is the study of materials. Materials are the "stuff" that things are made of.[1] We recognize the importance of the development and use of new materials in the history of humankind by identifying key periods in that history by the materials used, as in the Stone, Bronze, and Iron Ages. Frequently, the most exciting and important advances in materials science and engineering occur at the interfaces between, or by unconventional combinations of, traditional disciplines. This interdisciplinary research is carried out by scientists and engineers with training and backgrounds that include physics; chemistry; materials science and engineering (including the more traditional disciplines that focus on metallurgy, ceramics, and polymers); mathematics; electrical, chemical, civil, and mechanical engineering; and, increasingly, the biological sciences. Often, teams of researchers must be assembled to make progress on complex problems. This group process may occur in a "natural" way, following from the traditional modes of scientific exchange, or it may be induced by organization of the research environment through laboratory structure (typical of industry and some federally funded laboratories), geography (proximity of research groups, strategically placed common areas, and so on), and funding mode (group research programs of various types in several funding agencies). Collaborations may be formed around the conception or execution of research; different modes of collaboration are stimulated differently.

The first serious effort to induce group activity in academic materials research occurred when NSF assumed responsibility for the IDLs in 1972. Searching for some structure that would distinguish these block-funded, locally managed entities from the NSF-funded individual research on similar topics, NSF instituted the idea of Materials Research Laboratories (MRLs). MRLs consisted of a number of "thrust areas," each of which was to be focused on some broad problem requiring a multidisciplinary team of researchers. At this time NSF also created the overall materials management unit known as the Division of Materials Research (DMR).[2]

Focused research in areas of particular complexity that required a team of scientists in different disciplines became more and more common in the 1970s

[1]This observation has often been attributed to Paul Fleury, now dean of engineering at Yale University.

[2]For further reading about this period in the history of materials research, see National Research Council, *Materials Science and Engineering Through the 1990s*, Washington, D.C.: National Academy Press, 1989.

and was stimulated in part by the new culture engendered by the MRL program. These "seed" groups began to compete with other programs for funding. Until 1985, these groups could receive 3-year contracts from NSF after a lengthy evaluation process. To provide materials departments with a faster response to rapidly developing opportunities and developments within thrust areas, the NSF added the Materials Research Group (MRG) program after 1985. This program primarily targeted universities without an MRL, although some institutions with MRLs also received MRG funding. It is important to note that these two programs operated almost entirely independently.

The MRLs were deemed a success and used, in part, as the model for future NSF programs, including the Science and Technology Centers (STCs) and Engineering Research Centers (ERCs) developed in the 1980s, although these centers had different missions and operating structures. When DMR reorganized its group research program in 1994, it was natural to use the term "center" and dub these new entities "Materials Research Science and Engineering Centers." The research elements of MRSECs are organized into Interdisciplinary Research Groups (IRGs), with current centers composed of one to five IRGs. MRGs were eliminated as a separate program. As nanoscience and technology became more important, a new block-funded effort was developed and christened Nanoscale Science and Engineering Centers (NSECs).

These various types of NSF-funded centers differ in technical content. Some depend on internal group structure while others do not, and their management, duration, and funding levels vary. Centers do have elements of commonality; they are funded with the intention and mandate of carrying out activities beyond research. In the case of the MRSECs, they must manage shared experimental facilities (SEFs), conduct education and outreach (EO), interact with and transfer results to industry, and work toward a more diverse population of practitioners in the field of materials research. In addressing its charge, the committee examined each of these elements of the MRSECs, commencing with an introduction in Chapter 1, presenting the larger context of the program in Chapter 2, and then exploring the impact of research and facilities, education and outreach, and industrial collaboration in Chapters 3 through 5. Chapter 6 summarizes the committee's findings on the overall impact of the program and presents recommendations for restructuring group-based research in materials science and engineering at NSF.

MRSECs were created from the MRL program (and the much smaller MRG program) beginning in 1994, with all MRLs either terminated or converted to MRSECs by the end of 1996. Many new centers were created, for a total of 24 MRSECs at the end of 1996. At the same time, the budget for the MRL/MRSEC program increased from approximately $29 million per year (as-spent dollars) in 1993 to $44.28 million per year in 1996. This represented a change of 124 percent in the number of centers, but only a 53 percent increase in budgets. Clearly, MRSECs were "designed" to be smaller than MRLs, and some of the capabilities of the MRLs were reduced or eliminated in the transition. Most MRLs trimmed staff in shared experimental facilities and decreased the rate and value of equipment purchases for such facilities. More recently, the MRSEC as-spent budget slowly increased, and then essentially reached a plateau during the years 2003 to 2006 ($53.4 million for 2006).

From the outset of the MRL program, NSF managers and the research community have sought methods for evaluating the nature and quality of the work done in the locally managed, group-intensive laboratories (see the subsection entitled "MITRE Report" in Chapter 2). A study by the MITRE Corporation at NSF's behest in the late 1970s concluded that the research quality was comparable to that done by researchers not supported by the MRLs. The present committee sought to reexamine these questions in the context of this study of the MRSECs. The committee's overarching goal was not to specifically evaluate the MRSEC program, nor to recommend the continuation or termination of the program, but rather to describe and characterize its performance and impact and to make recommendations for the future of the program. The committee divided its analysis into several sections: research and facilities, education and outreach, and industrial interactions. These topics are addressed sequentially here; additional material can be found in each of the supporting chapters.

RESEARCH

In assessing the impact of the research enabled by the MRSEC program, the committee sought first to identify any unique, distinguishing features: Is the research enabled by the MRSEC program characteristically different from research enabled by other mechanisms? For instance, the charter of the MRSEC program refers often to the importance of collaborative, group-based research for advancing materials research. If the MRSEC program specifically enables group-based research, are the research results distinguishable from those developed by individual investigators? Or perhaps MRSECs enable research at a different phase of the overall progress in advancing the frontiers of materials science and engineering.

The committee found the task of evaluating the impact of MRSEC research quite daunting, primarily because research papers published in peer-reviewed journals rarely attribute the results to a single support mechanism. Moreover, any research, even by an individual researcher associated with a MRSEC, is a combination of activities supported "inside" and "outside" the MRSEC. Thus, even if MRSECs have played a unique role in the research enterprise, such as in enabling the formulation of research projects that could not otherwise have been envisioned, there is no easy way to provide substantiation.

Conclusion: Consistent with previous analyses, the committee found no simple, quantitative, objective measure to clearly differentiate the MRSEC research product from that of other mechanisms supporting materials science and engineering research.

Although the committee was unable to identify MRSEC-enabled research in "blind taste tests," it successfully assessed the overall research quality in comparison to the research enabled by other mechanisms and elsewhere around the world. For instance, it addressed the question, Do published research results that acknowledge MRSEC resources achieve citation indices and other measures of impact comparable to research enabled by individual-investigator awards?

The committee studied a set of major breakthroughs in materials research over the past four decades. U.S. universities, and in particular MRSECs and their predecessors the MRLs, played a limited but pivotal role in several of these discoveries. The committee conducted several comprehensive analyses comparing citations of MRSEC-associated research publications and those of the broader research community. The distribution of MRSEC-associated "top cited papers" across subfields of materials research was very similar to that of the top 100 most-cited papers. Affiliations of the top 100 research papers also showed a 10 percent contribution from institutions with MRSECs or MRLs. The committee also found that the top MRSEC papers were cited much more often than the average materials research paper, but that the best-of-the-best materials research papers had significantly more citations. However, these papers generally predate the emergence of the MRSEC program. The committee also found that the MRSEC program has the same level of collaboration as found in comparable national and international groups. To some extent, having fostered this type of research at an early stage may be the ultimate success of the MRSEC program. Finally, the departmental affiliations of MRSEC-associated authors and those of the top-cited materials research papers were quite similar.

In two related exercises, the committee examined the global stature of MRSEC-related research groups. In comparison to those for the Max Planck research institutes of Germany, the MRSECs' publication citation rates were quite comparable. In a peer-voting exercise, the committee contacted researchers around the world in several different subfields and solicited their opinions about world-leading research teams. Research teams at institutions with MRSECs dominated the results.

Although many of these measures are indicators of correlation and not causation, the committee came to believe that the research program enabled by MRSEC awards has been, in general, at least as effective as that enabled by other mechanisms.

Conclusion: Overall, the MRSEC program produces excellent, frontier science of the same high standard as that supported by NSF through other

mechanisms. In terms of quality, MRSEC research is at least on a par with that of other multiple-principal-investigator programs and individual grants in the United States and internationally, and is an important element of the overall mix for support of materials research, including support for big centers and single-investigator grants.

Since most publications acknowledge multiple sponsors, it is not possible to prove that MRSEC funding yields leadership in discoveries, publications, or citations in materials research. The lack of objectively quantifiable differences in research productivity or impact suggests that the unique value of the MRSEC program is in its broader impact to the local and national materials communities.

One could additionally wonder about the potential for a "chicken-and-egg" problem. At a strong institution with a MRSEC award, which came first, the strong campus research effort or the center? In the committee's judgment, the competitive selection process for MRSEC awards puts the burden on the pre-existing strength of the institutions. While a MRSEC award may enhance an institution's materials research programs, it will not necessarily bring them into being.

The committee's analysis led to several related general findings.

General Finding: Sponsors of research are increasingly unable to claim "sole ownership" of research results; MRSECs are no exception.

Most research publications now acknowledge multiple sponsors. It is not possible to demonstrate that the MRSEC support yields leadership in discoveries, publications, or citations. In part this is because funding per MRSEC has decreased significantly in the past decade, so that each group requires multiple sponsors.

General Finding: Most highly cited publications contain one or two senior authors, indicating that the size of research collaborations is usually small.

Although the materials field is highly collaborative and the general belief is that the community benefits from interactions among local groups of many individual investigators in the same field, discoveries and publication records indicate that over 50 percent of the published papers are from individuals and groups of two.

The committee notes that analyses of publications and citations are only sensitive to how the research work is carried out; it is much more difficult to determine how the research topics are conceived and what factors influence that process.

EXPERIMENTAL FACILITIES

In 2004, NSF's Division of Materials Research estimated that 12 percent of the MRSEC budgets was spent on capital equipment (typically from the IRG, Seed, and Facilities categories). The facilities budget also supports (at least in part) technical staff members, who train students and maintain the equipment. About $240,000 per year per MRSEC (on average) is spent on capital equipment. By rough estimate, about half of the equipment purchased through the NSF instrumentation programs (DMR's Instrumentation for Materials Research program or NSF's agency-wide Major Research Instrumentation program) within DMR ends up in a MRSEC facility. Through the MRSEC program, another $5 million (or an average of about $200,000 per center) is added to this amount. Assuming a 10-year life for forefront materials characterization equipment, a center might thus afford a total inventory of equipment of about $4.4 million.

The variations in actual capital spending from one MRSEC to another are considerable. The recent National Research Council report on shared experimental facilities (*Midsize Facilities: The Infrastructure for Materials Research*[3]) found that most SEFs that serve the large majority of the materials community have a $1 million to $50 million replacement capital value, with an average of about $10 million. At present, other sources of support for SEF equipment (typically, the universities themselves or, in some cases, foundations) are not large enough to make up the difference in needed support. Thus, the average age of equipment in SEFs continues to increase, with many individual items more than 20 to 25 years old.

Conclusion: The MRSEC program offers one of the principal opportunities in materials research to support shared experimental facilities (SEFs) that include not only equipment but also the personnel to provide training for students and to perform maintenance. Growing constraints on the per capita MRSEC budget have greatly diminished this ability, which is a concern for the infrastructure of materials research in general.

EDUCATION AND OUTREACH

Education and outreach (EO) covers a broad range of activities that serve audiences including K-12 students and teachers; undergraduate, graduate, and postdoctoral researchers; policy makers; and the general public. Consistent with the breadth of activities, EO projects serve many different purposes: educating future scientists and engineers; broadening the participation of underrepresented groups in science, technology, engineering, and medicine (STEM) disciplines; increasing science literacy among members of the public; informing the public about scientific and technical issues; improving K-12 science education; and enabling the development of a scientific and technical workforce.

[3]National Research Council, *Midsize Facilities: The Infrastructure for Materials Research*, Washington, D.C.: The National Academies Press, 2006, p. 304.

Although all NSF proposals are required to address the "Broader Impacts" of the proposed research,[4] an EO component is specifically required by the MRSEC program. Many (although not all) MRSECs have at least a part-time person (the EO coordinator) dedicated to managing EO projects. NSF does not require that specific activities or audiences be targeted by the MRSEC, with the exception of the Research Experiences for Undergraduates (REU) program, and a general dictum to broaden participation by underrepresented groups in STEM fields. MRSECs are encouraged to pursue activities consistent with the research and organizational/partnership opportunities of the center, as well as the size and local context of each center.

As with research, most MRSECs leverage their core EO funds with supplements and cooperate with other campus activities, making it difficult to separate the impact of the MRSEC per se. Although highly variable, about 10 percent of the total MRSEC budget is spent on EO activities and coordination, with much of this effort going to REU programs. The Research Experiences for Teachers (RET) program common to most MRSECs is funded from a program element at NSF located outside DMR.

Conclusion: The MRSEC education and outreach program has impacts on the NSF mission to educate and prepare the nation's future workforce.

- MRSECs provide unique opportunities for interdisciplinary research experiences that are different from those an individual student would experience in a single-investigator laboratory.
- MRSECs foster environments that support interactions with other programs to leverage funds and coordinate activities across campuses and disciplines. This culture leaves a vital imprint on students who work in MRSECs.
- MRSECs foster a mind-set of outreach and a sense of responsibility in current and future researchers.
- The centralized EO infrastructure that a MRSEC offers empowers researchers to engage in EO who would not ordinarily have done so.

General Finding: The most significant and well-documented contribution of MRSEC EO programs is the preparation of future researchers at all levels.

Research-related education and outreach activities leverage MRSEC strengths and expertise. MRSECs can provide unique opportunities for interdisciplinary

[4]See National Science Foundation, "Merit Review Broader Impacts Criterion: Representative Activities," 2002, available at http://www.nsf.gov/pubs/2002/nsf022/bicexamples.pdf.

research experiences that are different from those that an individual would experience in a single-investigator laboratory. Although broadening participation by women and underrepresented groups remains a challenge, the greatest contributions to meeting this challenge often come from EO programs such as REU and RET.

Conclusion: Although the committee's impression is that most MRSECs are doing good to excellent jobs with their EO programs and that many of these programs have a significant impact on their audiences, the lack of data to support these assertions poses a serious problem for NSF as it seeks to make the most efficient use of its resources.

REU and RET programs are much more likely to be evaluated than any other education efforts, although the evaluations focus primarily on logistics and self-reported participant perceptions. The quality of evaluations of other EO components varies greatly. MRSECs are reviewed primarily on the breadth of activities and the number of participants and not on documented outcomes.

General Finding: The future impact of MRSEC EO activities is threatened. The continued lack of specificity in EO expectations at the agency level has led to an emphasis on quantity over quality and innovation over impact.

General Finding: Most MRSECs feel compelled to participate in many disparate EO activities. This approach often does not make optimal use of the MRSECs' strengths, dilutes their potential impact, and in fact reduces the likelihood of determining what that impact is.

There is a perception that the demands of the EO program have grown significantly since the original inception of the MRSEC program. While the requests for proposals for the program show most growth in demands, the broad portfolio of activities, even in the smallest MRSECs, suggests that MRSEC resources are being spread too thinly and that the impact of those resources is being diminished. The committee did observe that although MRSEC per capita financial resources decreased over the past decade, the reported number of students involved has been growing. This trend suggests that non-full-time-equivalence is being used and that a greater variety of students are being exposed to MRSECs.

This perception should not be taken to suggest that the community does not value EO: the overwhelming majority of MRSEC participants expressed a belief that EO is important and enthusiastically participate in EO activities. Nevertheless, there is a strong belief among the MRSEC participants and prospective participants that the selection process rewards quantity over quality and innovation over impact. Two specific examples were mentioned most often:

- The belief that a MRSEC must reach all audiences, including K-12, undergraduate and graduate students, and the public; and
- The belief that continuing an existing, successful program is less well received than proposing something new.

The emphasis on breadth has led to evaluation that consists primarily of counting numbers of attendees, because the programs are so diffuse that more meaningful evaluation is impossible without funding from other sources. Some programs focus on generic outreach that has little to do with the MRSEC focus, much less materials science and engineering. While this type of outreach is important, it does not leverage MRSEC resources.

While current MRSECs mentioned that renewal reviews value doing something new over continuing programs that have been shown to be effective, the larger question is whether MRSECs should be required to innovate in the EO component of their programs, or whether the focus should be on using best practices to make an impact on their communities.

Focusing MRSEC resources into a select number of programs that address the local strengths and needs makes much more sense than trying to reach all audiences. The MRSECs that are successful in reaching a variety of audiences often are those with significant external funding for EO.

Recommendation: Education and outreach should continue to be part of the overall MRSEC portfolio; however, MRSECs should focus resources on programs with proven high impact that leverage each MRSEC's unique research strengths and that can be meaningfully evaluated.

The committee believes that EO is an important part of the MRSEC program but that steps can be taken to increase its effectiveness. In particular:

- MRSECs should focus on a limited number of activities that are aligned with MRSEC research goals, are consistent with the MRSEC size, leverage participant expertise and interest, and address local needs.
- Because of their documented impact, REU programs should continue to be required; providing research opportunities for faculty and students at predominantly undergraduate and minority-serving institutions should be strongly encouraged.
- MRSECs that offer RETs should provide teachers with research experiences in materials science and engineering. The RET is not meant to be primarily a curriculum-development program.
- Other EO projects should be peer reviewed by materials research education experts during the MRSEC proposal/review process. The best of these projects should be funded as long as the overall MRSEC is funded.

The RET recommendation is tempered by the committee's concern that the impact of the RET program is largely undocumented. The RET program is NSF-wide, so the lack of data is not solely a MRSEC issue. Cooperative efforts to document the impact of the program, as has been done with the REU program, are necessary. However, validating the program is beyond the scope of what should be expected as part of a MRSEC EO component.

Recommendation: In the context of the above recommendation, NSF should develop and support the MRSEC education and outreach community in sharing and facilitating ideas and resources, including best practices, for all activities. This would be especially helpful in the area of increasing the participation of underrepresented minorities.

A shift in emphasis from innovation to impact would make it easier for MRSECs to share best practices in EO. This would facilitate the distribution of EO materials already developed and decrease local re-invention of existing EO materials.

The Partnerships for Research and Education in Materials (PREM) program is an excellent example of how NSF can act as a catalyst for activities that involve women and underrepresented minorities in materials science and engineering research. The committee believes that centralized activities like PREM have a much higher probability of success than leaving each MRSEC to its own resources. NSF should leverage the experience of its MRSECs to identify and share successful strategies in this area not just with other MRSECs, but also with the materials science and engineering community as a whole.

Recommendation: NSF should provide appropriate guidance to MRSEC applicants and reviewers in order to refocus education and outreach activities and ensure the program's effectiveness.

It is evident to the committee that there is a multiplicity of EO activities in the MRSEC program, and that the lack of guidance from NSF to the MRSECs and reviewers has contributed to what appears to have become a less productive enterprise than it could be. This should not be so. Reviewers should receive clear instructions about the role of EO in the MRSEC: the impact of a MRSEC's EO program should be of cardinal importance. Further, MRSEC EO programs have different objectives and therefore should not be evaluated using the same standards as those for research. NSF funds educational research under other programs, and major initiatives should be supported through those programs, with a separate review system.

INDUSTRIAL INTERACTIONS

An important goal throughout the history of the MRSEC program has been to promote "active cooperation with industry to stimulate and facilitate knowledge transfer among the participants and strengthen the links between university-based research and its application," according to the program solicitation. Industrial outreach includes relevant sectors involved with the application of materials research beyond just commercial industries. Consequently, "industrial outreach" includes national laboratories and other federal entities (e.g., Department of Defense laboratories) that apply the results of basic materials research to address important national needs. MRSECs are required to develop and execute a program for knowledge transfer to industry. The MRSEC solicitation makes clear that this implementation should be flexible and consistent with the size, capabilities, mission, and vision of each individual MRSEC.

Industrial interaction may have direct benefits for MRSEC research programs that are stimulated by the challenges and research needs articulated by industrial partners. This positive feedback to the research planning was affirmed in discussions with numerous MRSEC directors. While responding to industrial challenges, MRSECs have maintained an appropriate focus on leading-edge and transformational research. To date, MRSEC industrial outreach appears to have been aimed primarily at large industrial research laboratories, but the opportunity to interact more with innovative small and start-up companies is increasing.

Conclusion: The program goals for MRSEC industrial collaborations are appropriate. A flexible approach to meeting those goals is essential to address the needs and capabilities of the individual MRSECs.

Conclusion: The MRSEC program requirement for industrial collaboration leads to important activities that likely would not occur otherwise (e.g., workshops, short courses, external advisory boards with industrial advisers).

The MRSEC directors whom the committee informally interviewed all were supportive of the industrial outreach and knowledge-transfer goals for the program. Although some centers had an existing campus culture that already supported industrial outreach activities, other MRSECs had to create a culture of industrial outreach to respond to program requirements. As a result, all centers had substantial outreach efforts that added significant value to the overall program. The committee found that local flexibility in meeting the program goals was effective in taking advantage of inherent differences among MRSECs, the university environment they resided in, and the targeted industrial community. As with education and outreach, there is a disproportionate impact on small centers to demonstrate accomplishments in all MRSEC program goals.

Conclusion: MRSECs have developed industrially relevant programs while maintaining a commitment to solving long-term research problems.

Maintaining this approach is important to the quality of the research efforts and to educational continuity for students, especially those involved in Ph.D. research programs. Industrial interactions are a positive part of the educational experience for students. The ability to connect their research to external needs and to have an opportunity to work with industrial scientists was clearly cited as a strength by students interviewed by the committee.

Conclusion: MRSEC industrial collaboration efforts are generally supported by multiple sources, in addition to MRSEC funds, such as funds from industrial partners themselves.

In a few cases, a significant portion of the MRSEC funding (more than 8 percent) was used for industrial outreach. More typically, MRSEC industrial outreach is supported primarily by university and/or state funding and is usually assisted by a university liaison program. This leveraging is valuable to the MRSEC program in meeting its goals, but it makes assessing the effectiveness of the industrial outreach program more difficult to judge as a function of MRSEC resources supporting the effort.

Conclusion: The importance given industrial collaboration and technology transfer in the review process is seen as not being commensurate with the importance of this program goal.

Each MRSEC tends to have its own program for industrial outreach and collaboration, and industrial contacts typically do not interact with more than one MRSEC. There is evidence of occasional industrial interactions that incorporate more than one MRSEC, but collaborative efforts among centers are the exception.

MRSEC leaders understand the change in the research landscape within the United States and are trying to respond appropriately. In particular, there is a shift away from a system dominated by several large, comprehensive industrial research laboratories toward a greater number of small and entrepreneurial companies involved with technology innovation. Understanding how to work effectively with these smaller companies and ensuring that these interactions are properly recognized and valued by the MRSEC program will be critical.

The committee was generally impressed with the breadth of the industrial outreach efforts across the MRSEC program. Each center seems to have a vital industrial outreach activity that meets the stated program goals. While it is difficult to clearly evaluate the impact of the industrial outreach efforts, the committee believes that the MRSEC program is generally meeting its goals and that the industrial outreach is valuable.

Recommendation: NSF should establish metrics for evaluating the effectiveness of industrial collaboration and technology transfer.

In addition to considering worldwide best practices, NSF should quantify the relative importance of industrial outreach and knowledge transfer relative to other program requirements in program solicitations. This would enable centers to put the appropriate focus and resources on this aspect of their center and would enable reviewers to make appropriate judgments about accomplishments.

Recommendation: Together with the team of MRSEC directors, NSF should provide a mechanism to enable industry to effectively understand the resources and expertise available through the network of MRSECs. This may require a coordination function that currently does not seem to exist, such as a national network liaison officer based at NSF.

Industrial outreach and knowledge-transfer effort is inherently based on interactions among people. Encouraging more personnel exchanges, such as student internships, extended sabbaticals for industrial researchers at MRSECs, visits by MRSEC faculty to key industry partners, significant industrial involvement on MRSEC advisory boards, and so on, will be essential to effective knowledge transfer and skill development (especially for students). The most common barrier to successful industrial interactions is simply a lack of contact among the relevant players. Taken together, the MRSECs represent a significant body of talents, tools, and expertise. The committee believes that better leveraging of this combined value could enhance industrial collaborations and technology transfer. For instance, a program liaison could centrally receive and guide inquiries and requests from potential industrial partners.

PERCEIVED AND MEASURED IMPACT OF MRSECs

Why do outstanding people and institutions pursue MRSEC grants with all of the associated responsibilities? Analysis of inquiries made of faculty at both MRSEC and non-MRSEC institutions revealed multiple motivations for participation in the MRSEC program.

Conclusion: MRSEC awards continue to be in great demand. The intense competition for them within the community indicates a strong perceived value. These motivations include:

- **The ability to pursue interdisciplinary, collaborative research;**
- **The resources to provide an interdisciplinary training experience for the future scientific and technical workforce from undergraduate to postdoctoral researchers;**
- **Block funding at levels that enable more rapid response to new ideas, and that support higher-risk projects, than is possible with single-investigator grants;**
- **The leverage and motivation MRSECs provide in producing increased institutional, local, and/or state support for materials research;**
- **The perceived distinction that the presence of a MRSEC gives to the materials research enterprise of an institution, thus attracting more quality students and junior faculty; and**
- **The infrastructure that MRSECs can provide to organize and manage facilities and educational and industrial outreach.**

These factors suggest that there are strong positive influences of the MRSEC program on the conception of research ideas and the ability to pursue them quickly and effectively, which in turn have clear, positive implications for maintaining and advancing U.S. research competitiveness in the materials field. This observation must be tempered in the context of the current funding situation, in which MRSECs are asked to take on increasing responsibilities without the availability of commensurate resources.

> **Conclusion: The committee examined the performance and impact of MRSEC activities over the past decade in the areas of research, facilities, education and outreach, and industrial collaboration and technology transfer. The MRSEC program has had important impacts of the same high standard of quality as those of other multi-investigator or individual-investigator programs. Although the committee was largely unable to attribute observed impacts uniquely to the MRSEC program, MRSECs generally mobilize efforts that would not have occurred otherwise.**

MRSECs conduct and publish research with characteristics similar to those of other programs. The shared-facilities element of MRSECs represents a significant portion of the NSF investment in midsize facilities for materials research. The MRSEC education and outreach programs clearly benefit from the sharing and pooling of resources; improvements by NSF and the participating communities are needed, however. Although industrial collaborations that take place within the MRSEC framework are similar in character to those elsewhere, the activities initiated by MRSECs generally represent efforts that would not have occurred otherwise.

AT THE BREAKING POINT?

The committee examined funding data supplied by NSF that characterized as-spent dollars for various programs and activities in DMR from 1996 to 2006. Support for the individual-investigator programs has increased by 34 percent in this period (although it has been decreasing slightly in the past 3 years), national user facilities by 45 percent, and instrumentation (Instrumentation for Materials Research and Major Research Instrumentation, although the latter is non-DMR funds) by 42 percent. The MRSEC part of the centers program has increased in this period by only 20.5 percent.

In 2006, the MRSEC budget of $53.48 million supported 26 active MRSECs and 3 MRSECs in phase-out funding. The average MRSEC budget is thus close to $2 million per year (with an actual range of $1.0 million to $3.8 million per year). The MRSEC budget is divided into six principal categories: IRGs (63%); Seeds (for rapid response to new ideas) (10%); Education and Outreach (10%); Shared Experimental Facilities (11%); Industrial Outreach (2%); and Administration (4%). As with the individual MRSEC total budgets, there is considerable variability from center to center in these categories. Individual MRSECs also leverage these funds through institutional commitments, user fees for shared experimental facilities, and/or industrial and state support.

An "average NSF budget" for a current MRSEC can be determined from these figures:

Category	Average Annual MRSEC Spending
Interdisciplinary Research Groups	$1,260,000
Seeds	200,000
Education and Outreach	200,000
Facilities	220,000
Industrial Outreach	40,000
Administration	80,000
Total	$2 million

Compounded by the decrease in spending power estimated using an approximate but realistic university inflation index developed by the committee in the subsection entitled "NSF and the Division of Materials Research" in Chapter 2, the average MRSEC can now undertake only about 70 percent of the effort that it undertook in 1996, and only 40 percent of the effort that an MRL could undertake in 1993. It is in this context of diminished resources that the committee examined the current program that not only consists of the original tasks of research and shared experimental facilities but now also includes education and outreach, diversity, and industrial interaction. More information about the origin of the MRSEC program and its historical role in materials research is presented in Chapters 2 and 3. Does this suggest that increased funding for MRSECs should be sought by decreasing other elements in DMR—for instance, the individual grants?

Analysis reveals that single investigators at DMR have faced similar conditions of attrition in purchasing power. From 1996 to 2005, the median DMR single-investigator grant increased from $83,786 to $112,333 in as-spent dollars, an increase of 34 percent. During this time the number of grants increased from 377 to a high of 561 and then decreased to 365 in order to increase the average size of the grants. While the size of the grants has increased in as-spent and even in Office of Management and Budget (OMB)–inflated dollars, it has decreased compared to university inflation and is much less than the overall increases in the NSF budget. This strain on the individual investigator is at least in part a consequence of the significant decline in DMR funding relative to other elements of the Mathematics and Physical Sciences (MPS) budget. It is unlikely and highly undesirable to address weakness in MRSEC funding by eroding further the already stressed individual-investigator grant program.

What then about seeking additional resources from elsewhere within NSF? According to NSF data, the NSF budget for research and related activities (uncorrected for inflation) increased from $2.046 billion to $4.333 billion from 1993 to 2006 (or an increase of 112 percent, a number that is substantially above university inflation). The situation for DMR is dismal by comparison: from 1993 to 2006, the budget increased from $175.3 million to $242.9 million (or by 38 percent, somewhat more than the OMB inflation index and well below the university inflation index). It is clear from these observations that DMR is losing the battle within NSF for its share of new resources. This committee was not charged to nor did it attempt to determine whether the issue is one of new program responsibilities for NSF or of waning success in convincing senior leadership of the continuing value of materials research and the needs within DMR.

It is clear that a major problem looms as prospects for the next decade of materials research funding at NSF are contemplated. Another decade of similar decreases will undermine the ability of the MRSEC program to make valuable contributions in the future.

Conclusion: The effectiveness of MRSECs has been reduced in recent years as a result of increasing requirements without a commensurate increase in resources. Increasing the mean grant size is necessary to allow the program to fulfill its important mission goals.

Average funding for centers, in constant dollars, has decreased substantially in the past decade. Declining funding has been particularly detrimental to building and maintaining the advanced instrumentation necessary for leading-edge materials research. Additional pressures have arisen from increasing industrial and education and outreach responsibilities per center coupled with an increasing number of MRSECs, while the MRSEC program has remained at a relatively constant budget level. As materials research has blossomed as a robust and stable enterprise, the MRSECs have been expected to handle more and more responsibilities for the community (facilitating education and outreach activities, promoting diversity, engaging industry in technology-transfer activities, acquiring and maintaining instrumentation and facilities, and so on). This trend is not sustainable.

MOVING FORWARD

Entry into the MRSEC program is highly sought. More than 100 preproposals were submitted in the last competition, which ended with only two new MRSECs added to the program. Few NSF programs can identify higher relative proposal pressure or smaller success ratios. The disappearance of the MRG program from DMR effectively relegates support for interdisciplinary group research to IRGs in centers only. This proposal pressure adds weight to the committee's conclusion that the MRSEC program is a valuable component of the U.S. materials research portfolio and should be funded and managed accordingly.

Conclusion: The MRSEC program needs to evolve in order to successfully meet its objectives in the coming decade. To do so, the National Science Foundation must restructure the program to reduce requirements, reduce the number of MRSEC awards, and/or increase the total funding of the MRSEC program while preserving its positive elements.

The MRSEC program is at a critical point in its history. The current trends suggest that, if the program is left unchanged, the capacities and competencies of the centers will be subject to both relative and absolute decline. Without an increase in total funding and/or a restructuring of the sort that the committee proposes, MRSECs will have to be smaller, operating research programs that have a more limited reach than those they replaced in the original Materials Research Laboratory system. To the extent that facilities cannot be supported, they will likely fail to rise either to state-of-the-art levels or to the standards being set by global competitors. Continuation of these trends suggests a program that will not be able to make significant or uniquely identifiable contributions to the national portfolio of materials research. It will be one of a class of programs that, in very similar ways, supports multi-investigator efforts at modest levels, albeit doing so with considerable overhead in the form of other requirements for service to non-research programmatic goals.

The committee's deliberations took place in the context of a national discussion about the future of U.S. global leadership in science, technology, and innovation that has been unfolding over the past few years. In October 2005, echoing widespread concerns, the National Academies' report *Rising Above the Gathering Storm*[5] outlined a program designed to enhance the U.S. science and technology enterprise so that the nation can sustain its cultural vitality, continue to provide leadership, and successfully compete, prosper, and be secure in an increasingly interconnected world. In particular, the report identified basic research in engineering and the physical sciences as a key underpinning of the nation's strategic strengths. Response to this call to arms has been strong in the current administration (which proposed significant additional funding for NSF, the Department of Energy, and the National Institute of Standards and Technology as a component of its American Competitiveness Initiative) and in both chambers of Congress where several bills have been approved in committee.

In the event that additional resources can be made available, the committee emphasizes the need to increase unit funding of MRSECs rather than increasing their total number, while also addressing the issues of program management that would enhance discipline-wide education and industrial outreach. Simultaneously, the committee would endorse the reestablishment of a Materials Research Group program to support those small-group efforts that now fall into the abyss between individual-investigator and large center efforts. If additional resources do not become available, the number of MRSECs would have to be decreased to achieve these goals.

There have been calls for renewed investment in the physical sciences and engineering (e.g., *Rising Above the Gathering Storm*[6]) as well as thoughtful discussion of the level of resources necessary to achieve the scientific goals in condensed-matter and materials physics (e.g., *Condensed-Matter and Materials Physics: The Science of the World Around Us*[7]). The committee firmly believes that the MRSEC program is an important and strategic investment in NSF's portfolio of materials research activities; however, the level of support is suboptimal. Additional resources and the restructuring indicated above could produce significant additional value.

Born from the MRL program, the MRSEC program represented the next step in an evolutionary process for centers-based research in materials. Since that time, the character of the research community has continued to evolve. Fully equipped centers play an important role in the enterprise, serving as nucleation points for

[5]National Academy of Sciences, National Academy of Engineering, and Institute of Medicine, *Rising Above the Gathering Storm: Energizing and Employing America for a Brighter Economic Future*, Washington, D.C.: The National Academies Press, 2007.

[6]National Academy of Sciences, National Academy of Engineering, and Institute of Medicine, *Rising Above the Gathering Storm: Energizing and Employing America for a Brighter Economic Future*, Washington, D.C.: The National Academies Press, 2007.

[7]National Research Council, *Condensed-Matter and Materials Physics: The Science of the World Around Us: An Interim Report*, Washington, D.C.: The National Academies Press, 2006.

facilities, outreach efforts, and even research planning activities such as workshops. Small teams of researchers, taking advantage of these centers and other resources, have become just as important. Trying to address both of these needs with one program with a standard element (the MRSEC) has begun to strain the program.

Recommendation: To respond to changes in the budgetary landscape and changes in the nature of materials research in the coming decade, NSF should restructure the MRSEC program to allow more efficient use and leveraging of resources. The new program should fully invest in centers of excellence as well as in stand-alone teams of researchers.

Resources for basic research, especially in materials research, have not kept pace with overall economic growth in the past decade. Expectations for the range and extent of impacts enabled by NSF's programs have also changed. And materials research has continued to mature as a discipline. The MRSEC program can be positioned to better facilitate research advances in the next decade by improving the focus of its resources on targeted, specific objectives and by increasing its flexibility to allow specialization for the strengths of individual centers. The committee developed one detailed vision for achieving these objectives; it is articulated here. The committee envisioned a transition to this new formulation of the program to be initially revenue-neutral.

Two related funding mechanisms could be created, under the auspices of the NSF Division of Materials Research: one (the Materials Centers of Excellence, or MCE, program) would support several coordinated teams of interdisciplinary research groups, carry out educational and industrial outreach, and support state-of-the-art facilities. The second would support interdisciplinary Materials Research Groups (MRGs) that do not have separately mandated educational and industrial activities or facilities. The rationale for this shift is to centralize the value-added activities at appropriately funded structures, without losing the benefits of the interdisciplinary research being done by smaller groups of researchers. The MCEs would take on more of the educational and industrial outreach and facilities development and maintenance responsibilities on behalf of the entire materials research community.

The committee notes a critical element in this proposal: a review process that compares and competes the research activities across the entire program. That is, the barrier between MCEs and MRGs should be permeable in both directions as well as outside the program. For instance, the new MCEs would be much like the present MRSECs (three to six research groups but of more flexible sizes) and with enhanced capabilities for "seed" research, equipment, types of outreach, and an explicit facility responsibility for the region. In review (for both renewal and entry into the program), the MCEs would be reviewed separately by committees as to the excellence of the science and as to the additional responsibilities of an

MCE. A successful MCE would demonstrate excellence in both areas and should be explicitly evaluated as greater than the sum of its parts. Additionally, the MRGs would be reviewed only on the excellence of the science. The reviews of the science at the MCEs and of the MRGs elsewhere should be done by experts in the particular subfields and be competitive. The reviews of the other aspects of the MCEs should be by experts in those areas. More information on the specifics is provided in Chapter 6.

DMR has mechanisms for collaborative, group-based research.[8] For instance, in 2006, there were 33 active Focused Research Group (FRG) awards that represented a total annual investment of about $11 million. Similarly, DMR supported 36 active Nanoscale Interdisciplinary Research Team (NIRT) awards in 2005 at a combined level of nearly $13 million. Although NIRTs are being phased out, renewal proposals are being directed to the Focused Research Group program. NIRTs are more like mini-centers, however. The committee draws an important distinction between the nature of research supported by these mechanisms and the chief characteristic of research enabled by the MRSEC program: the MRSEC program encourages collaboration in the conception of research, while the other programs facilitate collaboration in the execution of research. By providing intellectual and physical infrastructure up front, the MRSEC program encourages collaboration in the conception of research. The committee distinguishes the proposed MRG awards by their longer-term nature (5 or 6 years as opposed to 3 for FRGs). Finally, the committee's proposal envisions a direct and open competition among all the MRGs in a regular cycle.

There is tremendous opportunity to be realized if the MRSECs operate with greater cooperation and synergy. MRSECs largely conduct their industrial outreach programs completely independently of other MRSEC programs. There is evidence of occasional industrial interactions that incorporate more than one MRSEC, but collaborative efforts between centers are the exception. There could be a significant benefit realized if industry could effectively understand the resources and expertise available through the MRSEC program at the national level. This may require a coordination function that currently does not seem to exist, such as an overall national network liaison officer based at NSF.

Conclusion: NSF encourages MRSECs to operate as a national network. Although some efforts have been made in that direction, the committee did not observe strong cooperation among the discrete centers of the program. The

[8]According to the NSF Grants Program Guide, "A group proposal is one submitted by 3 or more investigators whose separate but related activities are combined into one administrative unit. A collaborative proposal is one in which investigators from two or more organizations wish to collaborate on a unified research project." Available at http://www.nsf.gov/funding/preparing/faq/faq_g.jsp?org=DMR#group; viewed May 1, 2007.

MRSEC program is thus missing a clear opportunity to leverage resources and thereby strengthen the materials research enterprise as a whole.

NSF has encouraged the individual MRSECs to work together as a network of centers that could enhance the program through cooperative effort. Annual meetings of MRSEC directors, as well as less frequent assemblies of education and outreach coordinators, have led to exchanges of best practices and shared concerns; however, there is little evidence of collaborative efforts stimulated by such interactions. Several MRSECs recently have started an NSF-funded effort to develop regional capabilities for shared facilities. This effort is to be commended, but there should be more efforts of this type.

Recommendation: NSF should enable its materials research centers to play a greater role in advancing materials research.

As centers for teams of investigators, MRSECs could play a natural role in facilitating community formulation of initiatives in materials research. Such activities might include but not be limited to organizing conferences and workshops addressing significant questions in materials research, creating and maintaining a national directory of MRSEC expertise and facilities, leveraging economies of scale in industrial and/or educational outreach, and providing geographically based infrastructure for materials research facilities.

OUTLOOK

The committee's analysis shows the MRSEC program to have had important impact over the past decade, about commensurate with that of the individual-investigators program within DMR. By virtue of the intense competition within the community for these centers, the committee concludes that they are perceived to be quite valuable. The chief feature of MRSECs that appears to be unique is their ability to create an environment of group-based research with sufficient scope and resources to foster interdisciplinary research and training of students. Similarly, MRSECs serve as resource centers for carrying out certain "broader impact" types of activities as part of NSF's mission.

Looking forward, the formulation of the MRSEC program needs to evolve to take advantage of a new generation of scientific progress and discovery. Group-based research has become an established element of the DMR portfolio, and the MRSEC program should focus on empowering small, nimble research groups as well as larger infrastructure nodes with their own competitive research teams. This evolution will help ensure NSF's position as a leading supporter of the world's most important materials research.

1

Introduction

Charged with assessing the impact of a specific program, the National Science Foundation's (NSF's) Materials Research Science and Engineering Centers program (MRSEC program), the MRSEC Impact Assessment Committee chose to examine that program in the context of its intended goals (see Box 1.1, entitled "The MRSEC Program Mission Statement") and the role of its field of materials research in the overall portfolio of federally funded research. Three elements of that overall portfolio most critical to the nation's health, prosperity, and security are the biological, information, and materials sciences. Of these three, materials science is the most complex to "manage," as it intersects and depends on most other disciplines, requires group as well as individual efforts, and is equipment-intensive at levels from small to medium scales. This chapter develops the background required to assess the role of the NSF MRSEC program in materials research, its effectiveness, and opportunities for improvement.

THE LANDSCAPE OF MATERIALS RESEARCH

The present era is a broadly diversified materials age. The many remarkable technologies that are now part of daily life are enabled by newly developed materials, including transistors and memory devices, artificial body parts that extend useful life for the physically impaired, high-strength concrete enabling modern construction, lightweight materials enabling air travel, and many, many more. How did these materials come to be available and where do we expect the next generation of materials to emerge from? The process of development and transition to market

BOX 1.1
The MRSEC Program Mission Statement

Following is the current mission statement of the Materials Research Science and Engineering Centers program (MRSEC program):

> MRSECs [Materials Research Science and Engineering Centers] are supported by NSF [National Science Foundation] to undertake materials research of a scope and complexity that would not be feasible under traditional funding of individual research projects. NSF support is intended to reinforce the base of individual investigator and small group research by providing the flexibility to address topics requiring an approach of broad scope and duration. MRSECs incorporate most or all of the following activities to an extent consistent with the size and vision of the Center:
>
> - Programs to stimulate interdisciplinary education and the development of human resources (including support for underrepresented groups) through cooperation and collaboration with other organizations and sectors, as well as within the host organization. Cooperative programs with organizations serving predominantly underrepresented groups in science and engineering are strongly encouraged.
> - Active cooperation with industry to stimulate and facilitate knowledge transfer among the participants and strengthen the links between university-based research and its application.
> - Cooperation and collaboration with other academic organizations and national laboratories.
> - Active efforts to establish research collaborations and education activities at the international level are strongly encouraged. Cooperative activities may include, but are not limited to: joint research programs; affiliate programs; joint development and use of shared experimental facilities; access to user facilities; visiting scientist programs; joint educational ventures; joint seminar series, colloquia or workshops.
> - Support for shared experimental facilities, properly staffed, equipped and maintained, and accessible to users from the Center, the participating organizations, and other organizations and sectors.
>
> Each MRSEC has the responsibility to manage and evaluate its own operation with respect to program administration, planning, content and direction.[1]

[1]National Science Foundation, *Program Solicitation for Materials Research Science and Engineering Centers*, NSF 04-580, Washington, D.C., 2004.

is a complex story, but underlying it is the materials research and development (R&D) supporting the invention and fabrication of such new materials.

Materials research has some features that differentiate it from other types of science and engineering. The work tends to be of a long-range character, and new materials tend to have far-reaching implications for many other fields of science, from medicine to high-energy physics, and for the economic and strategic health of the nation. In spite of the importance of materials research, there is a tendency to defer the difficult work of creating new materials to others. Since the payoff is often very remote from the enabling research, the impulse can be to concentrate research on immediate applications rather than on fundamental enabling science. While such a policy may appear attractive, concentrating on brief, short-term benefits to the detriment of long-term gains vastly undercuts future scientific capability. All fields of science share this feature, of course, to varying degrees.

Another common requirement for most experimental work in materials research is access to many different types of small- to medium-sized equipment. The variety of tools required for structure, composition, and properties characterization is far too extensive and expensive to be found in a single investigator's laboratory. Sharing equipment, either through informal means or through organized facilities, is a major component of carrying out the materials research endeavor.

It is useful to place the MRSEC program in the context of the overall field of materials research. The committee summarizes its views on the overall field in the following list of definitions:

- *Materials*—Perhaps the most useful and descriptive definition is that materials are "the stuff of which things are made." Invoking a now-traditional rubric, the committee recognizes the importance of the development and use of new materials in the history of humankind through the identification of key periods in that history, such as the Stone, Bronze, and Iron Ages, in terms of the materials that characterize them. The present era is a broadly diversified materials age. The technological wonders that are now part of daily life are enabled by the newly developed materials from which they are made. These developments include the transistors and memory devices that power computers, telephones, and high-definition televisions; the artificial body parts that extend useful life for the physically impaired; the high-strength concrete that enables modern construction; the lightweight materials that surround passengers in air travel; and much more. How did these materials come to be available for use by modern designers, and where do we expect the next generation of materials to emerge from? The process of development and transition to market is a complex story, but underlying it is the materials research and development supporting the invention and fabrication of such new materials.
- *Materials research*—The subject of the MRSEC program technical agenda is the study of materials. What does that mean? The most recent comprehensive study of this subject, made in the late 1980s by the National Research Council (see Box 1.2, "Materials Research in National Research Council and Other Reports"), defined materials science and engineering as having four integrated elements: synthesis/processing, structure/composition, proper-

BOX 1.2
Materials Research in National Research Council and Other Reports

In 1993, the National Research Council (NRC) issued the report *Science, Technology, and the Federal Government: National Goals for a New Era.*[1] In that report, the Committee on Science, Engineering, and Public Policy (COSEPUP) suggested that the United States adopt the principle of being among the world leaders in all major fields of science so that it could quickly apply and extend advances in science wherever they occur. In addition, the report recommended that the United States maintain clear leadership in fields that are tied to national objectives, that capture the imagination of society, or that have a multiplicative effect on other scientific advances. These recommendations were reiterated in another NRC report, *Allocating Federal Funds for Science and Technology*[2] (1995), which said that the United States should "strive for clear leadership in the most promising areas of science and technology and those deemed most important to our national goals."

In 1999, the National Science and Technology Council (NSTC) stated that advanced materials are the foundation and fabric of manufactured products.[3] To support its assertion, the NSTC cited the role of advanced materials in, among other uses, fuel-efficient automobiles, damage-resistant buildings and structures, electronic devices that transmit signals rapidly over long distances, the protection of surfaces from wear and corrosion, and the endowing of jet engines and airframes with sufficient strength and heat tolerance to permit ever-faster supersonic flight. The NSTC concluded that many leading commercial products and military systems could not exist without advanced materials and that many of the new products critical to the nation's continued prosperity would only come to be through the development and commercialization of advanced materials.

In its report *Experiments in International Benchmarking of US Research Fields* (2000),[4] COSEPUP asked how important it is for the United States to lead in materials science and engineering (MSE). The materials subpanel that wrote the MSE-focused sections of that report noted that there had been an explosion in the understanding and application of MSE since the end of World War II and that connections had become stronger between the materials field and other fields with emerging technology. The result, the subpanel concluded, was an acceleration in the contributions of materials to social advancement and economic growth.

The reports cited above represent only a small sample of the many volumes that have been produced on the importance of materials research to future U.S. economic and national security and how the United States

[1]National Academy of Sciences, National Academy of Engineering, and Institute of Medicine, *Science, Technology, and the Federal Government: National Goals for a New Era*, Washington, D.C.: National Academy Press, 1993.

[2]National Academy of Sciences, National Academy of Engineering, Institute of Medicine, and National Research Council, *Allocating Federal Funds for Science and Technology*, Washington, D.C.: National Academy Press, 1995.

[3]Office of Science and Technology Policy, National Science and Technology Council, *1998 Annual Report*, Washington, D.C., 1999, p. 24.

[4]National Academy of Sciences, National Academy of Engineering, and Institute of Medicine, *Experiments in International Benchmarking of US Research Fields*, Washington, D.C.: National Academy Press, 2000.

ties, and performance.[1] Research supporting any or all of these elements is a proper subject for materials research by individuals, groups, or centers. That research includes experiments, theory, and simulation and modeling.

[1]National Research Council, *Materials Science and Engineering for the 1990s: Maintaining Competitiveness in the Age of Materials*, Washington, D.C.: National Academy Press, 1989.

should react to the changing environment in which MSE research and development (R&D) are taking place. The numerous reports on the subject all point out that MSE research continues to address issues in agriculture, health, information and communication, infrastructure and construction, and transportation. Some areas of particular interest are these:

- The national defense of the country continues to depend on providing accessibility to the most advanced weapons to the military, and the evolving threat to homeland security demands new materials to solve new problems.
- MSE research continues to provide solutions to problems in health care with the development of new materials for the delivery of life-saving drugs and new implant technologies.
- MSE research is producing advanced materials solutions for more efficient energy-production and -transmission systems.
- MSE research is providing the latest materials for advanced transportation needs, such as for more energy-efficient and safer automobiles and advanced aerospace systems.
- Numerous consumer products benefit from MSE R&D.

Given the multifaceted importance of MSE R&D to the United States, maintaining world leadership in the field remains a critical national priority.[5] As described in the recent NRC report on the globalization of materials R&D,

> The discovery, understanding, and exploitation of new materials and phenomena are the heart of CMMP [condensed-matter and materials physics]. Invention and innovation in this field have had a pervasive impact on our daily lives. Examples are everywhere: semiconductor lasers are in our DVD players; advanced magnetic materials store data on our computers' hard drives; liquid-crystal displays show us our photographs and our telephone numbers. But these technological marvels tell only half the story: studies of new materials and phenomena have also led to significant advances in our basic understanding of the physical world. For example, the development of ultra-pure layered semiconductors made possible not only the production of high-speed transistors for cell phones, but also the discovery of completely unexpected new states of matter. Efforts to understand magnets, ferroelectrics, superconductors, polymers, and liquid crystals, exploited in innumerable applications, spurred the development of the elegant, unified conceptual framework of broken symmetry that not only explains how the characteristic behaviors of these materials are related, but also underlies much of modern physics. These examples illustrate the inherent intertwining of the pure and applied aspects of condensed-matter and materials physics; they are opposite sides of the same coin that define and enrich the field.[6]

[5]National Research Council, *Globalization of Materials R&D: Time for a National Strategy*, Washington, D.C.: The National Academies Press, 2005.

[6]National Research Council, *Condensed-Matter and Materials Physics: The Science of the World Around Us: An Interim Report*, Washington, D.C.: The National Academies Press, 2006, p. 1.

It is carried out at universities, in government laboratories, and within industry. It may involve single investigators or groups. It may be done in small laboratories or at huge facilities such as synchrotron, neutron, and high magnetic field sources. It may deal with fundamental underlying principles, the invention of new materials, the characterization of structure and properties, the development and refinement of processing (manufacturing), the prediction of in-service life expectancy, and even environmentally friendly disposal.

- *Materials researchers*—Materials research is carried out by scientists and engineers with training and background that includes physics; chemistry; materials science and engineering (including the more traditional disciplines that focus on metallurgy, ceramics, and polymers); mathematics; electrical, chemical, civil, and mechanical engineering; and, increasingly, the biological sciences.
- *Interdisciplinary nature*—Materials research is interdisciplinary by definition and by evidence of the diverse backgrounds of its practitioners. Advances in materials research depend on individuals and results associated with many traditional disciplines (see Box 1.3, entitled "Origins of the 1996 Nobel Prize in Physics in the Materials Research Laboratories"). Frequently the most exciting and important advances occur at the interfaces between traditional disciplines, forever altering the scope and boundaries of those disciplines.

BOX 1.3
Origins of the 1996 Nobel Prize in Physics in the Materials Research Laboratories

In 1957, Bardeen, Cooper, and Schrieffer published their theory of the microscopic origins of superconductivity. Two years later, Phil Anderson proposed that some variation on this theory might suggest that other degenerate Fermi fluids might show similar condensed states. Anderson predicted a superconducting transition temperature of about 80 millikelvin (mK) for superfluidity in helium-3 (^{3}He). However, by 1965, physicists had cooled ^{3}He at near its vapor pressure to 2 mK, and no superfluid phase transition was observed. After that, the international search for a Bardeen-Cooper-Schrieffer (BCS) superfluid ended. However, in the same year, Yu D. Anufriev, a member of Peter Kapitza's laboratory in Moscow, for the first time attempted to cool liquid ^{3}He through the adiabatic compression and solidification of some of the liquid. This improbable cooling technique, first proposed by Isaac Pomeranchuk in 1950, allowed Anufriev to cool his liquid sample from 80 mK to about 20 mK. A few people believed that this technique might ultimately allow one to cool the liquid so low in temperature that the solid formed would exhibit nuclear-spin ordering.

David Lee, at Cornell University, was one of these people. With support from the Cornell Materials Center (one of the National Science Foundation Materials Research Laboratories [MRLs]) for fundamental research in low-temperature materials physics, he hired Robert Richardson as a postdoctoral associate in order to study this technique. In the autumn of 1971, Douglas Osheroff, a graduate student of David Lee, while studying how his Pomeranchuk refrigerator worked, discovered a kink in a curve of the melting pressure in the cell versus time. This kink was found to be extremely reproducible, and Osheroff and his mentors realized that it was the signature of some highly reproducible phase transition within this mixture of liquid and solid ^{3}He. They labeled this as the "A" transition. They estimated the temperature to be about 2.6 mK, but the solid nuclear-spin-ordering

- *Often a multidisciplinary process*—One strategy for achieving these advances at the disciplinary interfaces depends on the rare individual who is able to move beyond traditional disciplinary boundaries into unexplored territory. Often, but not by any means exclusively, the research requires multidisciplinary action in order to proceed. In such instances, individuals from two or more traditional disciplines make critical impacts along the way to success. This may be done in sequence or in some sort of collaborative, parallel mode. This multidisciplinary process may occur naturally, following from the traditional modes of scientific exchange, or it may be induced by the organization of the research environment, including the laboratory structure, typical of industry and of some federally funded laboratories, and by funding through group research programs.

This important subject of materials research has of course been addressed in many reports, including some by the National Research Council, as cited in Box 1.2. There the committee notes several excerpts that reinforce the position that

transition was only expected to occur at 2.0 mK. Ultimately the signature of a second transition, a "B" transition at well below 2.0 mK, was also found. The group employed a crude form of magnetic resonance imaging to separate out the behavior of the liquid and solid ^{3}He. On April 20, 1972, at 2:40 a.m., Osheroff noticed that at the lower of these two transitions the magnetic susceptibility of the liquid dropped nearly discontinuously by more than a factor of two. He wrote in his lab notebook: "Have discovered the BCS transition in liquid ^{3}He tonight." However, the group still believed that the A transition was in the solid phase.

On June 4, 1972, David Lee convinced Osheroff to remove his magnetic field gradient to see if the nuclear magnetic resonance (NMR) frequency of the solid shifted below the A transition temperature. What the two saw was completely unexpected. The solid signal did not move, but the liquid signal shifted continuously to higher and higher frequencies, until they saw the pressure signature of the B transition, at which point the liquid signal disappeared as it moved back under the much larger solid signal. Clearly, both the A and B transitions were in the liquid, and the ordered liquid exhibited very strange NMR properties. A preprint of their results was sent to Anthony Leggett at the University of Sussex, and in less than a month Leggett showed how a p-wave BCS superfluid could exhibit the strange NMR frequency shift seen at Cornell. Ultimately, Lee, Osheroff, and Richardson shared the 1996 Nobel Prize for physics for their discovery, and Leggett shared the 2003 Nobel Prize for physics for his theory of these remarkable fluids.

These initial discoveries in basic research, fostered by the MRLs, had profound influences. To this day, the basic research materials program at Cornell is world-class. Inspired by the Nobel Prize–winning work with low-temperature fluids, Leggett became a major force in the accomplishments of the Materials Research Laboratory at the University of Illinois at Urbana-Champaign where he is stationed. This remarkable story of instrumentation, discovery, and scientific accomplishment was made possible by the MRL program with its multidisciplinary approach to the combination of physics, chemistry, and engineering that later became known as materials research.

careful attention to the management of this research is a critical responsibility of the government.

NATIONAL SCIENCE FOUNDATION

The National Science Foundation Act of 1950 (Public Law 81-507) set forth NSF's mission and purpose: "To promote the progress of science; to advance the national health, prosperity, and welfare; to secure the national defense. . . ." The act authorized and directed NSF to initiate and support the following:

- Basic scientific research and research fundamental to the engineering process,
- Programs to strengthen scientific and engineering research potential,
- Science and engineering education programs at all levels and in all the various fields of science and engineering,
- Programs that provide a source of information for policy formulation, and
- Other activities to promote these ends.

Over the years, NSF's statutory authority has been modified in a number of significant ways. In 1968, authority to support applied research was given by the Daddario-Kennedy Amendment (Public Law 90-407). In 1980, the Science and Engineering Equal Opportunities Act (Public Law 96-516) gave NSF standing authority to support activities to improve the participation of women and minorities in science and engineering. Another legislative amendment effecting a major change occurred in 1986, when engineering was accorded equal status with science. In official agency words, the modern vision for NSF is as follows:[2]

> The National Science Foundation is a catalyst for progress through investment in science, mathematics, and engineering. Guided by its longstanding commitment to the highest standards of excellence in the support of discovery and learning, NSF pledges to provide the stewardship necessary to sustain and strengthen the Nation's science, mathematics, and engineering capabilities and to promote the use of those capabilities in service to society.

As an element of the NSF portfolio in the Division of Materials Research, the MRSEC program is necessarily tasked to advance the frontiers of research in materials research science and engineering.

---

[2]National Science Foundation, "National Science Foundation Strategic Plan," http://www.nsf.gov/nsf/nsfpubs/straplan/vision.htm.

RESEARCH CENTERS

From a philosophical standpoint, the idea of a research center offers two chief advantages over the disaggregated efforts of a collection of individuals. First, by allowing the pooling of resources and efforts, a center could achieve more benefit either through economies of scale (e.g., simple efficiency arguments for equipment sharing) or by breaking through a critical-mass threshold. For instance, in terms of education and public outreach, one might imagine that coordinating the efforts of a dozen faculty in a MRSEC into a coherent approach (such as developing a regular relationship with a nearby secondary-school classroom) could be much more effective than a dozen different such ad hoc efforts. Second, by bringing people together from a variety of backgrounds, a center might foment intellectual synergy.[3,4] On a university campus, a center might offer additional benefits by allowing a set of like-minded faculty to speak with a single voice to the university administration, federal research agencies, or even other members of the research community.

It is important to note that no single strategy will be successful in the short and long term; a portfolio of approaches is required for a robust program of lasting value (e.g., both individual and center-based researchers will always be necessary).

NSF Research Centers

The first serious effort to induce group activity in academic research occurred when NSF assumed responsibility for the materials laboratories formerly known as Interdisciplinary Laboratories for the study of materials and run by the Advanced Research Projects Agency (ARPA). Searching for some structure that would distinguish these block-funded, locally managed entities from the individual research on similar topics funded by the Foundation, NSF instituted the idea of Materials Research Laboratories (MRLs) consisting of a number of "thrust groups," each of which was to be focused on some broad problem requiring a multidisciplinary team of researchers. Other groups of this type have been subsequently constituted by NSF in its Materials Research Groups and its Interdisciplinary Research Groups (a key element of the current MRSECs). NSF has extended this idea to other disciplines through its Focused Research Groups, and the concept is emulated by the Department of Defense (DOD) in its Multidisciplinary University Research Initia-

---

[3]National Academy of Sciences, National Academy of Engineering, and Institute of Medicine, *Facilitating Interdisciplinary Research*, Washington, D.C.: The National Academies Press, 2004, pp. 39, 189.

[4]National Research Council, *An Assessment of the National Science Foundation's Science and Technology Centers Program*, Washington, D.C.: National Academy Press, 1996, p. 20.

tive groups. The concept of group research is now a well-established element in academic circles and a particularly common one in the field of materials research.

Aggregations of scientists and engineers in large groups are often referred to as centers or laboratories. Within the academic environment, the term "center" is now most common, perhaps because of the history of the NSF funding. The Materials Research Laboratories within NSF were deemed a success and used, in part, as the model for future programs, including the Science and Technology Centers (STCs) and Engineering Research Centers (ERCs) that were developed in the 1980s. When the MRLs were reconstituted in 1994, it was natural to use the term "center" and dub them Materials Research Science and Engineering Centers (MRSECs). Similarly, as new block-funded efforts were developed in the burgeoning field of nanoscience and technology, they were named Nanoscale Science and Engineering Centers (NSECs).

The ERC and the STC programs differ largely because of their long-term award and the expectation that the centers will evolve toward being supported by other types of support at the end of the award. The ERCs are typically focused around a specific research problem that is likely to transition to a successful market need. Industrial partnerships are strongly encouraged, and at the end of the 10-year award (assuming successful renewal at the 5-year mark), the center could be supported entirely by industrial funds. STCs typically focus on basic research problems in multidisciplinary areas. Both ERC and STC awards are "sunsetted" after 10 years, because it is expected that at the end of the award the research problem will either have been solved or have been transitioned to another domain (such as systems engineering). NSF's NSEC program is more similar to the MRSEC program, although the 5-year award can be renewed only once. Because MRSECs focus on basic research topics, which differs from work at these other centers, they enjoy the opportunity to renew their awards competitively every 6 years.

These NSF-funded centers differ in technical content. Some depend on internal group structure while others do not, and their management, duration, and funding levels are quite varied. Centers do have elements of commonality: they are funded with the intention and mandate of carrying out activities in addition to the research that justifies their existence. In the case of the MRSECs, they must manage central research facilities, conduct education and outreach, interact with and transfer results to industry, and work toward a more diverse population of future practitioners in the field of materials research.

Through its work, the committee came to believe that centers in general and MRSECs in particular are "community builders." This sense is hard to quantify and objectively measure, of course, but easy to acquire on speaking with members of the communities. The center concept has been successful—certainly as judged by the enthusiastic participation and by the number of proposals from those seeking to participate—spawning many different types of centers at NSF: STCs, ERCs, NSECs, as well as dedicated user facilities (National High Magnetic Field Laboratory, Cornell High Energy Synchrotron Source, Synchrotron Radiation Center, and so on) and the smaller "group" efforts (such as Integrative Graduate Education and Research Traineeships, FRGs, and so on).

The program solicitation for MRSEC proposals has evolved since the first offering in 1993. The emphasis on international partnerships and collaborations is a recent addition, for instance. The committee therefore chose not to assess the performance and impact of this element of the program.

Materials research spans many different classical academic disciplines even at universities that include an explicit materials science department. These disciplines include applied physics, chemistry, chemical engineering, electrical engineering, mechanical engineering, physics, and others. While in principle individuals could "self-assemble" into broad, interdisciplinary groups to tackle important problems, there are few examples of that occurring in an academic setting. MRSECs (and now many of the other centers) encourage and enable broader interactions among faculty in these departments by providing joint funding for such activities.

The original Interdisciplinary Laboratory (IDL) concept of materials centers was motivated by perceived national needs in materials that were unlikely to be met by the "stovepipe" mentality that resulted from departmental and college organizational structures. IDLs were created as one of the earliest elements of the present-day Defense Advanced Research Projects Agency (DARPA), which itself was created in response to the Russian launching of Sputnik and a perceived weakness in U.S. research. IDLs were intended to dramatically increase the nation's research on materials, and the mode of funding was developed recognizing the superb models that existed in industry (especially Bell Laboratories and General Electric Laboratories) and that had been so successful during the Manhattan Project. Thus, if universities were to be strengthened in this area, they would need new resources, but they would also have to change the way they were performing research. By contrast, industrial R&D is rarely organized in ways that reflect academic disciplines, for good reason. Many of the problems tackled by industry (most especially in development activities, but also in research) require interaction and inputs from many disciplines as part of a team effort. Indeed, the general decline in industry-sponsored basic research has opened a significant gap in the nation's science and technology enterprise. University-based centers are attempting to bridge this gap by putting increased effort into connecting their research with industrial interests. For example, the MRSEC at the University of California, Santa Barbara, has major relationships with Mitsubishi Chemical and Air Products, each of which includes an explicitly negotiated intellectual property agreement and sponsorship of multiple graduate student and postdoctoral research projects.

Other Federal Research Centers

The MRSEC program is one of several NSF centers-based programs.[5] All have similar programmatic elements, with some differences in emphasis and organization. For example, the ERC program focuses on close collaboration and translational research with industry for use in end-applications of great variety. The STC program is similarly problem-driven and topically diverse, but it emphasizes large, multiple-entity collaboration. NSECs, like MRSECs, generally have a dominant MSE component and focus on the nanometer-length scales—a subject matter that could also be addressed via ERCs, STCs, and MRSECs. ERCs, STCs, and NSECs share a sunset clause that limits the existence of any particular center to approximately 10 years. The NSF fiscal year (FY) 2007 budget request to Congress describes the NSF portfolio of centers as shown in Table 1.1. To be clear, MRSECs do not comprise the total NSF investment in centers-based materials research; the research programs of the NSECs, created in 2001, overlap significantly with those of the MRSECs.

Research centers represent 4 to 5 percent of the overall NSF budget. The breakout in Table 1.1 suggests that MRSECs represent 22 percent of the "centers spending" at NSF and 31 percent of the number of centers; that is, individual MRSECs receive less support than that provided the average NSF center. Materials centers are also the oldest centers-based program at NSF, when considering the program's direct ancestors.

Table 1.2 suggests that MRSECs are, by comparison with other NSF center programs, "leveraged" in an above-average way and that, per NSF dollar spent, the number of participants is above average (100 participants per million dollars).

Selected Centers at NIH

The National Institutes of Health (NIH) requested about \$2.77 billion in FY 2007 for assorted research centers, or about 9 percent of the overall agency budget. The total number of research centers is cited at about 1,400, but of these, the 94 biotechnology centers are the most relevant subset. The biotechnology centers have an aggregate funding level of \$131 million, representing an average per center level of funding similar to that of the MRSEC program (29 centers, \$52 million). These NIH centers have five key elements: technological research and development, collaborative research, service work for researchers who are not part of a center, education and training, and dissemination of research results or techniques.

[5]Lists of institutions receiving support through the ERC, MRSEC, NSEC, and STC programs can be found at http://www.erc-assoc.org/, http://www.nsf.gov/funding/pgm_summ.jsp?pims_id=5295&from=fund, http://www.nsf.gov/funding/pgm_summ.jsp?pims_id=7169, and http://www.nsf.gov/od/oia/programs/stc/, respectively.

TABLE 1.1 National Science Foundation Research Centers Programs, Selected from the President's Budget Request for FY 2006

	Center Funding ($ millions)						
	Program Initiation (year)	Number of Centers, FY 2005	Budget, FY 2005	Budget, FY 2006 Current Plan	Budget, FY 2007 Request	Change over FY 2006 Budget (amount)	Change over FY 2006 Budget (percent)
Centers for Analysis and Synthesis	1995	2	7.07	6.39	6.46	0.07	1.1
Chemistry Centers	1998	6	3.00	1.48	3.00	1.52	102.7
Earthquake Engineering Research Centers	1988	3	6.00	6.00	—	−6.00	−100.0
Engineering Research Centers	1985	19	62.31	63.42	62.79	−0.63	−1.0
Materials Research Science and Engineering Centers	1994	29	52.41	53.66	55.70	2.04	3.8
Nanoscale Science and Engineering Centers	2001	15	36.40	37.21	37.35	0.14	0.4
Science and Technology Centers	1987	13	49.65	62.38	67.48	5.10	8.2
Science of Learning Centers	2003	4	19.83	22.71	27.00	4.29	18.9
Total		91	236.67	253.25	259.78	6.53	2.6

NOTE: Totals may not add due to rounding.
SOURCE: National Science Foundation, *FY 2006 Budget Request to Congress,* Washington, D.C., p. 419.

This multipronged mission has significant overlap with the expected roles of the MRSECs, although the NIH centers perhaps emphasize the relationship to the broader community more heavily.

Selected Centers at DOD

The Department of Defense, primarily through the research offices of the service branches and through ARPA/DARPA, has been one of the largest support-

TABLE 1.2 Levels of Participation in National Science Foundation (NSF) Centers-Based Programs in FY 2005

	FY 2005 Estimates for Selected Centers						
	Number of Participating Institutions[a]	Number of Partners[b]	Total NSF Support ($ millions)	Total Leveraged Support[c] ($ millions)	Number of Participants[d]	Leveraging Percentage	Participants per Million Dollars of NSF Support
Centers for Analysis and Synthesis	4	20	7	2	736	28.6	105.1
Chemistry Centers	53	19	3	4	269	133.3	89.7
Earthquake Engineering Research Centers	65	155	6	10	1,130	166.7	188.3
Engineering Research Centers	280	482	62	72	8,310	115.6	133.4
Materials Research Science and Engineering Centers	103	325	52	42	5,274	80.1	100.6
Nanoscale Science and Engineering Centers	130	269	36	16	1,630	44.0	44.8
Science and Technology Centers	94	306	50	28	2,118	56.4	42.7
Science of Learning Centers	20	11	20	8	366	40.3	18.5
Total	749	1,587	237	182	19,833	76.9	83.8

NOTE: Statistics reported for Science and Technology Centers are for 2004 only. Information is not yet available for new centers funded at the end of FY 2005.

[a]All academic institutions that participate in activities at the centers.
[b]Total number of nonacademic participants, including industry, states, and other federal agencies.
[c]Funding for centers from sources other than NSF.
[d]Total number of people who use center facilities, not just persons directly supported by NSF.

SOURCE: National Science Foundation, *FY 2006 Budget Request to Congress,* Washington, D.C., p. 425.

ers of materials research over the past 60 years. Generally, the DOD components have not funded infrastructure/facilities, with some notable exceptions. The most important exception for materials research came with the DARPA IDL program, which provided "user fees" that enabled universities to construct new buildings for the interdisciplinary materials research and which supplied the original capitalization that launched major characterization facilities at these universities.

In the early 1980s, DARPA made a major investment in facilities by establishing three gallium arsenide (GaAs) foundries for the development of GaAs device manufacturing processes. These foundries were given to the Rockwell Science Center, McDonald Douglas Company, and AT&T. The foundries had specific device goals set by their contract but did provide manufacturing services to the III-V community.[6] Also, the Defense University Research Instrumentation Program (DURIP) is designed to improve the capabilities of U.S. institutions of higher education to conduct research and to educate scientists and engineers in areas important to national defense by providing funds for the acquisition of research equipment. A central purpose of DURIP is to provide equipment to enhance research-related education. The last solicitation made 214 awards worth $43.5 million, averaging about $200,000 each.

The DOD supports centers-based materials research through the programs described below.

Multidisciplinary University Research Initiative The DOD Multidisciplinary University Research Initiative (MURI) is sponsored by the DOD research offices: the Office of Naval Research, the Army Research Office, and the Air Force Office of Scientific Research. The MURI program supports research in basic science and/or engineering that is of critical importance to national defense. The program is focused on multidisciplinary research efforts that intersect more than one traditional science and engineering discipline. More than half of the MURIs are materials-research-related.

By supporting individual multidisciplinary teams, the MURI program complements other DOD basic research programs that support university research through single-investigator awards. The total amount of funding for 5 years available for grants resulting from the FY 2005 program solicitation is estimated to be about $135 million, pending out-year appropriations. It is anticipated that the average award will be $1 million per year, with the funding for each award dependent on the scope of the proposed research. By contrast with the NSF MRSEC program, these MURIs do not require expenditures on equipment or outreach.

---

[6]The III-V notation refers to chemical compounds, typically metal oxide in nature, formed with elements from the third and fifth columns of the Periodic Table of the Elements.

University-Affiliated Research Centers The DOD University-Affiliated Research Center program creates research centers within universities for military applications. Examples of such centers are the Institute for Soldier Nanotechnologies at the Massachusetts Institute of Technology (MIT); the Institute for Collaborative Biotechnologies at the University of California at Santa Barbara, with MIT and the California Institute of Technology as subcontractors; and the Institute for Creative Technologies at the University of Southern California. These centers each receive about $10 million per year from the Army Research Office and focus on basic and applied research, including applied research collaborative with industry, with an emphasis, for example, on meeting soldier needs via new products for communication, situational awareness, personal protection, and energy supply.

LOOKING FORWARD

The MRSEC program is the latest stage in the evolutionary development of group research in materials funded by the National Science Foundation. The challenge faced by this study committee was to examine the health of this program after more than a decade in the present mode and to suggest opportunities for improvements as NSF contemplates the next stage in this evolution.

2

The Overall Context of the Materials Research Science and Engineering Centers Program

The very complexity of material interactions means that much of the research tends to require extensive experimental trial and error—that is, an Edisonian approach that takes into account the latest results from theoretical analysis and computational modeling. This approach has no guarantee of success, it could require many years, and as in most scientific endeavors, most trials do fail. The benefit of broad-based, long-term efforts across many subfields of materials science is the only way to ensure a healthy, continuous rate of scientific accomplishment. This model is one that has traditionally been supported by the federal government to complement science in general, including the materials science research conducted in academic venues.

The need for brute-force trial-and-error investigations is partially mitigated by the availability of sophisticated analytical instruments. These instruments often allow researchers to obtain profound insights because such tools shed light on the underlying physical principles that govern phenomena; other tools allow researchers the ability to precisely synthesize or construct systems of interest. In order to pool resources and optimize the utilization of these complex and often quite expensive instruments, the tools are frequently collected in a central facility that provides expert staff, maintenance, and training. As a result, long-term financial-support mechanisms are needed to cover not only the initial capital investment (often millions of dollars) but also the ongoing resources needed to enable them to operate to their full capacity over the long term. Thus, the two main ingredients for a successful materials research enterprise are, first, patient, long-term support and, second, a large array of expensive analytical, synthetic, and processing equipment.[1]

Both of these requirements can only be met by long-term, patient funding that is sufficiently centralized to support a full suite of the most advanced analytical and synthetic instruments. Patient research support, combined with major centralized instrumentation, was the formula for the original Materials Research Laboratory (MRL) program that was the precursor of the current Materials Research Science and Engineering Centers (MRSECs). In fact, the 1999 National Research Council report *Condensed Matter and Materials Physics: Basic Research for Tomorrow's Technology* stated: "New facilities and instrumentation create new opportunities in condensed-matter and materials physics, and continued support for facilities and for broad access to them must be emphasized."[2]

The current guidelines for competition for MRSEC funding have had two effects.[3] The size of the average MRSEC award has shrunk, and the funding has been divided into smaller increments that are too small to adequately support the needed analytical and synthetic centralized facilities. As the infrastructure of instrumentation and facilities is subsequently eroded, the scientific benefits of those centers are thereby diminished. The second penalty is that the constant competition for and turnover of the smaller MRSECs prevents the long-range risk taking that is part of the nature of successful materials research. As noted in *Midsize Facilities: The Infrastructure for Materials Research:*

> The committee recognizes a . . . need for midsize facilities that have . . . sufficient size and complexity, either in instrumentation or in the supporting technical staffing or even building infrastructure, to require that significant attention and resources be spent on supporting these core activities. The committee terms these core activities "long-term infrastructure" and recognizes that, as required at the larger national facilities, steady funding and stewardship are required to make midsize facilities work more effectively over the long run.[4]

In this field, some fraction of the funding must be highly stable in order to allow major risk taking. This risk taking would occur most naturally in the context of a center that is large enough to accommodate both near-term efforts and the

[1]National Research Council, *Midsize Facilities: The Infrastructure for Materials Research*, Washington, D.C.: The National Academies Press, 2006, pp. 3, 38, 78-80.

[2]National Research Council, *Condensed-Matter and Materials Physics: Basic Research for Tomorrow's Technology*, Washington, D.C.: National Academy Press, 1999, p. 304.

[3]The committee gathered information about the operational history of the MRSECs through testimony at meetings, phone interviews, data prepared by the National Science Foundation, and by reviewing the series of program solicitations.

[4]National Research Council, *Midsize Facilities: The Infrastructure for Materials Research*, Washington, D.C.: The National Academies Press, 2006, p. 134.

long-term programs. The long-term programs would permit an adequate emphasis on broad-based materials exploration and development.

The MRSECs exist in an interesting culture of interdisciplinary and multidisciplinary research, one that has come to characterize much of materials research. It is on this "cutting edge" that MRSEC research is supposed to exist. The direction of research at any institution at a given time is set by factors such as budget, organization, current trends, and perceptions of needs. While this environment can and has led to many amazing breakthroughs, materials research is currently in a time of constrained or decreasing budgets. At the same time, there is an increased concern about how lagging technical leadership retards the economic competitiveness of the U.S. economy. Great opportunity lies at the interdisciplinary frontiers that MRSEC research explores.

SCIENTIFIC CONTEXT

MRSECs are supported by the National Science Foundation (NSF) to undertake materials research of a scope and complexity that would not be feasible under traditional funding of individual research projects. The research focus at an individual MRSEC is divided along the lines of the Interdisciplinary Research Groups (IRGs)—research groups of varying size—which do not necessarily have commonality with one another, even within the same center. This structure is meant to provide a vehicle for achieving the center's research mission, which follows from NSF's mission.

Recently, there has been a trend in the materials research community toward addressing "grand challenges" of materials research.[5,6] Given the mission and structure of the MRSEC program, the centers are encouraged by NSF to conduct such "transformative" research.

As an exercise, the MRSEC Impact Assessment Committee developed a list of grand challenges for materials research—energy, health care, water purification, information technology, national security, and so on—in addition to "hot" technologies that could result from materials research. This exercise was meant only for instructional purposes since the subject matter is beyond the scope of this study.

The interim report from the Committee on CMMP (Condensed-Matter and Materials Physics) 2010 addresses the question of important challenges in a more unifying way, focusing on what the committee sees as the broadest research issues

[5]National Research Council, *Condensed-Matter and Materials Physics: Basic Research for Tomorrow's Technology*, Washington, D.C.: National Academy Press, 1999, p. 304.

[6]National Research Council, *Condensed-Matter and Materials Physics: The Science of the World Around Us: An Interim Report*, Washington, D.C.: The National Academies Press, 2006.

of both scientific and technological interest related to materials.[7] These include emergent properties and complexity, energy, physics of life, matter far from equilibrium, nanoscale phenomena, and advanced measurement and prediction. Defining the substance of the materials research frontiers is not the subject of the current report, but it is abundantly clear from even this brief initial discussion of grand challenges and hot technologies that there is a huge variety of issues that require research activities based at major centers to be part of the overall approach.

HISTORICAL CONTEXT

The MRSEC program is descended from a long history of federal investment in institutions designed to promote and support materials research as part of the nation's research enterprise. Because of the important context set by the history of the program (and its evolution), the important predecessors of the MRSEC program are briefly noted here.

History

The MRSEC program is a descendant of the Interdisciplinary Laboratories (IDLs) begun by the Advanced Research Projects Agency (ARPA, later the Defense Advanced Research Projects Agency [DARPA]) under the Department of Defense (DOD) in 1960 (see Figure 2.1 for a history of the program). The IDL program was intended to support interdisciplinary research in materials science, mainly for application to military uses. Although obvious changes and transitions have been made in U.S. materials science programs since then, it is evident that the MRSEC program's current ambitions do reflect its origins.

The United States began its formal investment in materials science with the overarching National Materials Program, generated by President Dwight Eisenhower through the White House Office of Science and Technology and the Science Advisory Committee in 1958-1959. Partly because of its unique ability to manage 5-year grants for research (lengthier than others), DOD took on oversight of this new initiative in 1959. The program was assigned internally to ARPA, which named the funding program and its new facilities the Interdisciplinary Laboratories. The work statement from the original ARPA IDL contracts stated:

> The Contractor shall establish an interdisciplinary research program and shall furnish the necessary personnel and facilities for the conduct of research in the science of materials with the objective of furthering the understanding of the factors which

[7] National Research Council, *Condensed-Matter and Materials Physics: The Science of the World Around Us: An Interim Report*, Washington, D.C.: The National Academies Press, 2006.

influence the properties of materials and the fundamental relationships which exist between composition and structure and the behavior of materials.[8]

It is important to note that in this early era of materials research, few universities contained academic departments of a sufficiently broad nature to be named "materials science" departments (see Table 2.1).[9]

The first three IDLs (at Cornell University, the University of Pennsylvania, and Northwestern University) were established after a competition by ARPA. The original three were followed a few years later by additional ARPA contracts, three from the Atomic Energy Commission (AEC; later incorporated in part into the Department of Energy), at the University of California at Berkeley, the University of Illinois at Urbana-Champaign, Iowa State University, and two from NASA.

At the peak of the IDL funding in 1969, these laboratories supported 600 faculty members and 2,385 graduate students and produced 360 Ph.D.s. The research efforts of all those involved in the IDLs were grouped into 134 "work units," separately characterized by particular research thrusts. These work units, however, lacked a focused team approach, which would later be fostered by the MRL program, upon transfer of the IDL program to NSF.

The IDL program garnered much success, but during the late 1960s the DOD began to reevaluate its role in basic, "non-mission-oriented" research at universities, and after a thorough program review in 1971, the IDL program was transferred to NSF and renamed the Materials Research Laboratory (MRL) program in 1972 (see Table 2.2). At the time, it was perceived that NSF was the chief option for the transferring of the IDLs from DOD. This move was mandated in FY 1968 by the Mansfield Amendment to a DOD spending bill that forced DOD to divest itself of research not directly related to its mission.

Once transferred to NSF, MRL grants became block funding grants rather than a group of principal-investigator (PI) awards operating under an umbrella award, as was true under the IDL program. By encouraging actual team collaboration between faculty in neighboring departments, this change enabled a more collaborative team approach than was possible under the DOD IDL program.

However, the transition was not without its own organizational challenges and disruptions. The NSF responded to the challenges with the creation of the Division of Materials Research (DMR), into which were integrated some of the more traditional materials programs in areas of physics and chemistry.

Focused research in areas of particular complexity that required a team approach of several scientists in different disciplines became more and more common

[8]Work statement from Advanced Research Projects Agency Interdisciplinary Laboratory program contracts, 1960.

[9]National Research Council, *Advancing Materials Research*, Washington, D.C.: National Academy Press, 1987.

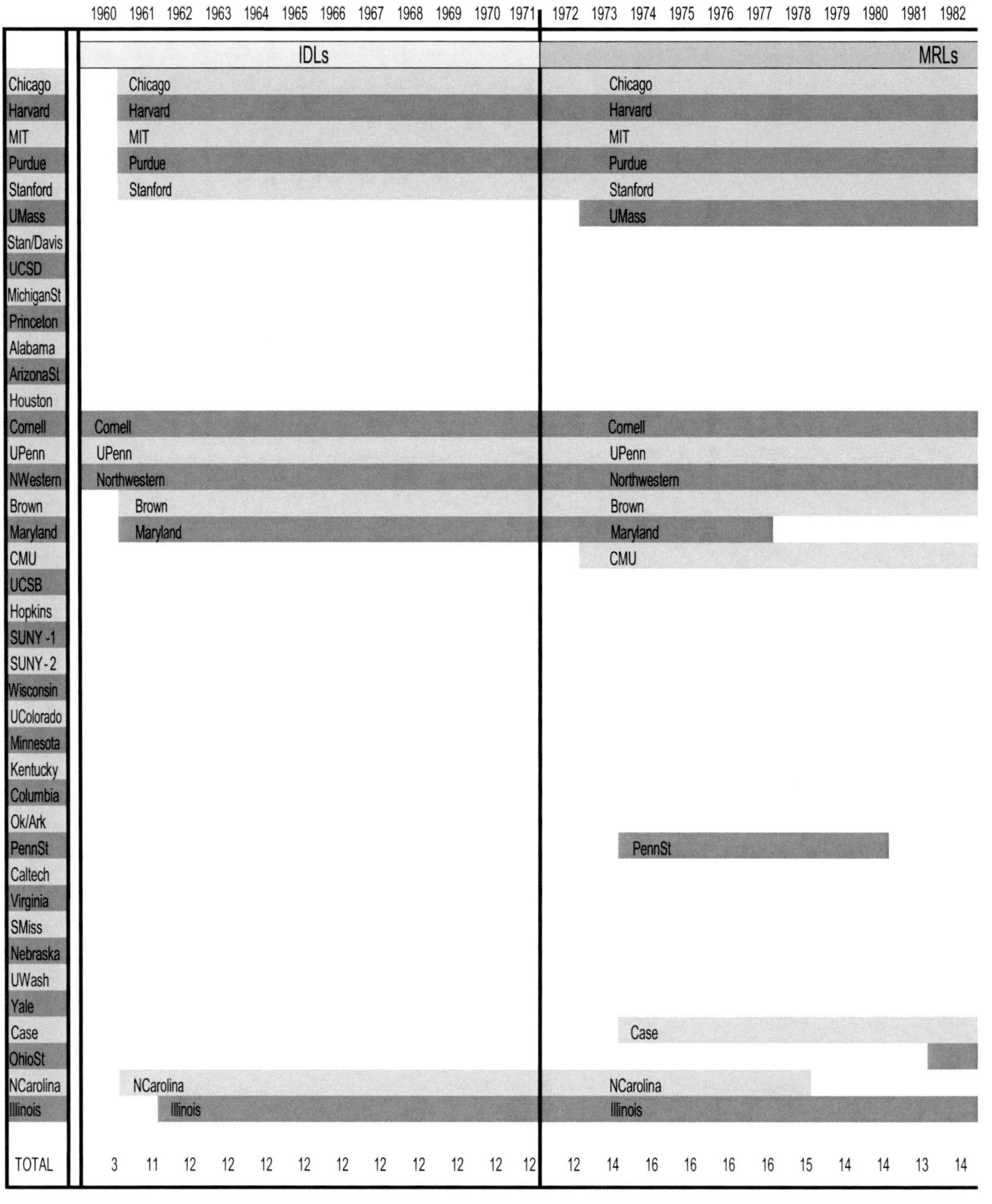

FIGURE 2.1 A center-based historical time line of the Interdisciplinary Laboratories (IDLs), Materials Research Laboratories (MRLs), and Materials Research Science and Engineering Centers (MRSECs).

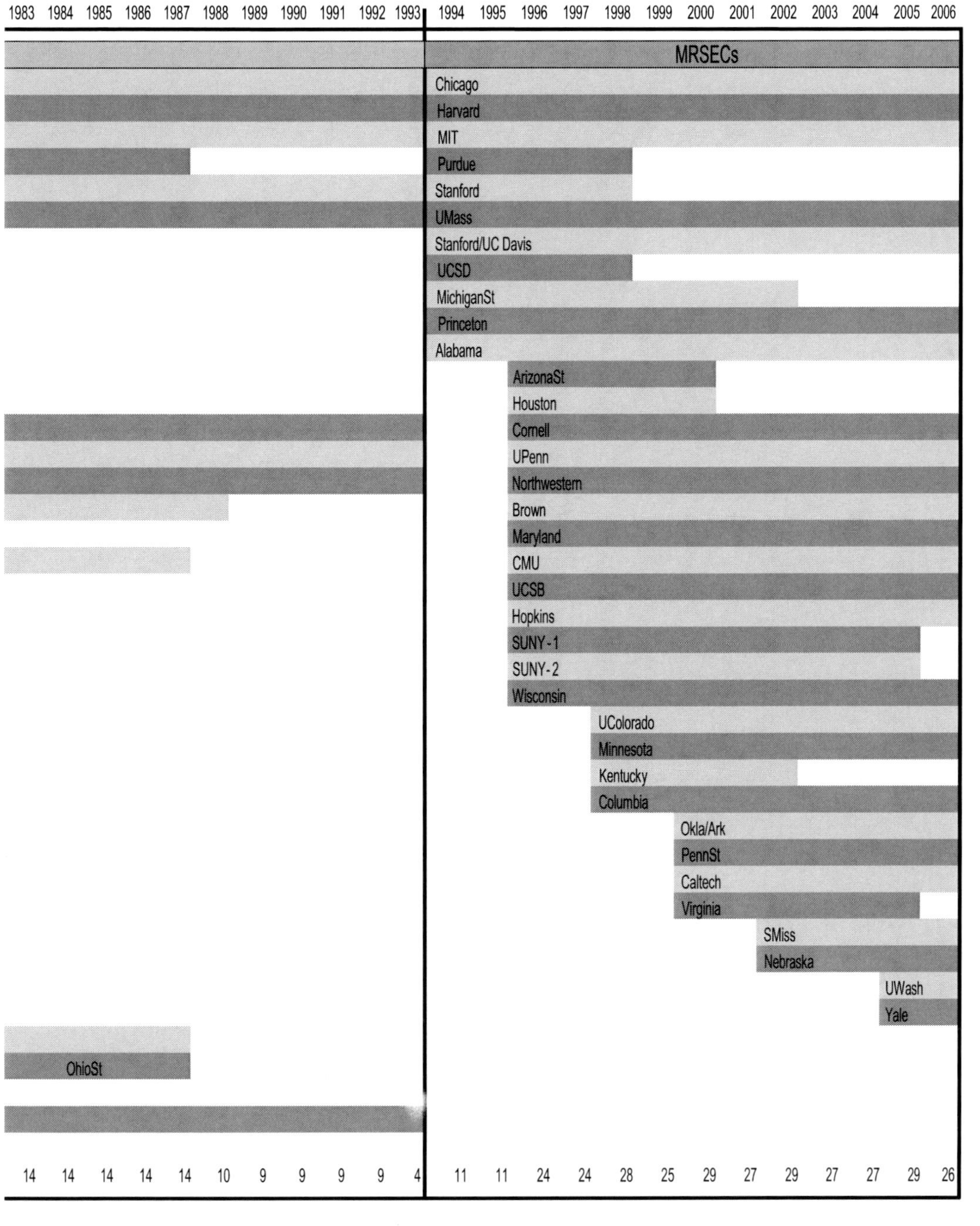
1983 1984 1985 1986 1987 1988 1989 1990 1991 1992 1993 1994 1995 1996 1997 1998 1999 2000 2001 2002 2003 2004 2005 2006
MRSECs
Chicago
Harvard
MIT
Purdue
Stanford
UMass
Stanford/UC Davis
UCSD
MichiganSt
Princeton
Alabama
ArizonaSt
Houston
Cornell
UPenn
Northwestern
Brown
Maryland
CMU
UCSB
Hopkins
SUNY-1
SUNY-2
Wisconsin
UColorado
Minnesota
Kentucky
Columbia
Okla/Ark
PennSt
Caltech
Virginia
SMiss
Nebraska
UWash
Yale
OhioSt
14 14 14 14 14 10 9 9 9 9 4 11 11 24 24 28 25 29 27 29 27 27 29 26

TABLE 2.1 Trends in Titles of Materials Research Academic Departments at U.S. Universities, 1964-1985

Department Title	Number of Departments, by Year		
	1964[a]	1970[b]	1985[b]
Minerals and Mining	9	7	5
Metallurgy	31	21	17
Materials	11	29	51
Other	18	21	17
Total	69	78	90

[a]Compiled from 1964-1970 *ASM Metallurgy/Materials Education Yearbook*, ed., J.P. Nielsen (American Society for Metals, Metals Park, Ohio).

[b]Compiled from 1985 *ASM Metallurgy/Materials Education Yearbook*, ed., K. Mukherjee (American Society for Metals, Metals Park, Ohio).

in the 1970s, owing to the new culture engendered by the MRL program. Funding for these "seed" groups began to compete with other programs for funding. Until 1985, these groups could receive only 3-year contracts from NSF, after a lengthy evaluation process. To provide materials departments with fleeter response to rapidly developing opportunities and developments within thrust groups, the NSF added another program, the Materials Research Groups (MRGs). This program

TABLE 2.2 Year of Establishment and Termination of Interdisciplinary Laboratories (IDLs) and Materials Research Laboratories (MRLs)

IDL/MRL University	Year Initiated	Year Terminated
Cornell	1960	1993
Pennsylvania	1960	1993
Northwestern	1960	1993
Brown	1961	1988
Chicago	1961	1993
Harvard	1961	1993
Maryland	1961	1977
Massachusetts Institute of Technology	1961	1993
North Carolina	1961	1978
Purdue	1961	1987
Stanford	1961	1993
Illinois (Urbana)	1962 (with AEC)	1993
Carnegie Mellon	1973	1987
Massachusetts (Amherst)	1973	1993
Pennsylvania State	1974	1980
Case Western Reserve	1974	1987
Ohio State	1982	1987

NOTES: Materials Research Groups were formed at the time of the phase-out of the MRLs at Carnegie Mellon University, Case Western Reserve University, Purdue University, the University of Michigan, Michigan State University, and the University of Texas at Austin. AEC, Atomic Energy Commission.

SOURCE: National Science Foundation and National Research Council, *Advancing Materials Research*, Washington D.C.: National Academy Press, 1987.

primarily targeted funding universities that did not have an MRL; however, some MRLs also received MRG funding. Table 2.2 lists the establishment and termination dates of IDLs and MRLs at institutions between 1960 and 1987.

Political trends in the late 1980s and early 1990s moved toward better maneuvering the nation's science investment to impact the economy through technological progress and educational outreach. For instance, George A. Keyworth II, President Ronald Reagan's Science Advisor and director of the Office of Science and Technology Policy, referred to the Engineering Research Centers organized by the NSF as "the single most important thing that we've done as an Administration in increasing the efficiency and effectiveness of federal R&D dollars."[10]

In 1992, a commission of the National Science Board (NSB) wrote a letter report in which it stated that research in the industrial sector was becoming more sharply focused on market-related issues, with fewer companies supporting long-term research.[11] The report recommended that the NSB and NSF should encourage interdisciplinary work and cooperation among sectors, and that NSF should encourage further development of joint science, engineering, and management education programs.

In response to these pressures, NSF reorganized its interdisciplinary MRL and MRG programs into the current MRSEC program. In the shift, the program began to focus on several aspects that its predecessors did not.

MITRE Report

As one component of adjusting its management style to the newly acquired materials research laboratories, in 1976 NSF asked the MITRE Corporation to conduct a study to:

- Analyze the effectiveness of Materials Research Laboratory (MRL)-type funding as a mechanism for the support of basic research in the materials science area;
- Identify the characteristics of MRL-type funding that may be appropriate for research support in other areas of research in science or technology; and
- Be useful for documenting oversight and program accountability, for planning program improvements, and as a model for evaluating similar federal research programs.

---

[10] *Science and Government Report* 15(18): 4 (1985).

[11] National Science Foundation, National Science Board Commission on the Future of the National Science Foundation, "A Foundation for the 21st Century: A Progressive Framework for the National Science Foundation," Washington, D.C., November 20, 1992, p. 4.

In 1978, MITRE published its report, *Evaluative Study of the Materials Research Laboratory Program.*[12] The report surveyed 16 MRLs, including 3 centers that had phased out or were in the process of doing so, and it constructed a comparison control group from the top 15 universities in project grant funds from DMR. The evaluation also included two Department of Energy (DOE) and two NASA laboratories sponsored under their IDL programs.

The MITRE study had conducted extensive peer review of 690 research papers "selected by statistical sampling techniques from MRLs and project-funded institutions." Citation analysis was then undertaken on more than 2,000 published papers. Data on other factors, such as equipment inventories, major research achievements, and number of doctoral degrees were also collected. The major research achievements, submitted by the MRLs, were reviewed by a panel of 19 experts.

Most notably, the MITRE study concluded:[13]

- Universities with MRLs have a better capability (in terms of faculty and equipment) to perform materials research than non-MRL universities without non-NSF materials science centers. The capability of non-MRL universities with materials science centers with funding from non-NSF sources is much like those with MRLs.
- About 70 percent of the materials research conducted at the MRLs was "unique" as compared to other research supported by NSF and undertaken at those institutions.
- There are no significant differences between universities with and without MRLs in concentration of funding, annual rate of turnover in research areas, duration of research areas, and continuity of staffing.
- The review of research publications does not show a clear-cut dominance of one population over the other being compared. There is no statistically significant difference at the 90 percent confidence level among any of the populations with respect to interdisciplinarity and overall indicators of innovation. In quality of procedures, the NSF core-funded papers rank higher than project-funded ones. In contributions per paper to research or technology, NSF core-funded papers rank lower than project-funded. In the use of essential specialized equipment, excluding computers, core-funded papers rank higher than papers from universities without MRLs but with non-NSF materials science centers, but lower than papers from DOE/IDLs.
- Citation analysis shows that only NSF/Project-funded papers at MRLs were cited with significantly greater frequency than MRL core-funded papers. The latter were cited with about the same frequency as papers from DOE/IDLs and NSF/non-MRLs without Materials Science Centers.
- In major achievements, the MRLs have much more than a proportional share (based on total NSF funding) rated in the top 15 percent. However, the MRLs

[12]J.T. Ling et al., *Evaluative Study of the Materials Research Laboratory Program, Summary Report,* MTR 7764, McLean, Va.: The MITRE Corporation, 1978.

[13]J.T. Ling et al., *Evaluative Study of the Materials Research Laboratory Program, Summary Report,* MTR 7764, McLean, Va.: The MITRE Corporation, 1978, pp. iv-v.

have slightly less than a proportional share of achievements rated in the top 25 percent.

Overall, the MITRE report found that research conducted by MRLs is not more integrated or interdisciplinary in nature than research conducted by the NSF project grants. However, the report concluded that MRLs were "sole contributors" to specialized areas of research such as high-risk research.

An earlier study conducted by the National Academy of Sciences in 1974-1975, entitled *Materials and Man's Needs*,[14] analyzed whether the achievements of block funding at materials centers could have been possible had the faculty instead been funded directly. Of particular interest to this committee, the report indicates the following:

- There is little or no correlation between magnitude of block funding and development of the institution as a "materials school."
- There is only modest correlation between the availability of block funding and the existence of specialized laboratory buildings, or central facilities, or their scale.
- There is no correlation between large block grants and degree of interdisciplinary interaction.

The current report returns to these same questions with some more recent information in Chapter 3.

BUDGET CONTEXT

To fully understand the impact of the MRSEC program, the committee found it necessary to compare the scale of effort undertaken by MRSECs to the broader context of materials research. Levels of investment are one metric for doing so.

National Investments

The U.S. federal government has supported basic research in materials since the post-World War II era (see Figure 2.2).

Although the committee could not find distinct data illustrating the history of industrial support for materials research performed in the academic sector, Figure 2.3 shows that for research in general performed by academic institutions, industry's contribution (now about $2 billion annually) has remained a small fraction of the federal level. Actual industry funding in inflation-adjusted dollars

[14]National Academy of Sciences, *Materials and Man's Needs*, Washington, D.C.: National Academy Press, 1975.

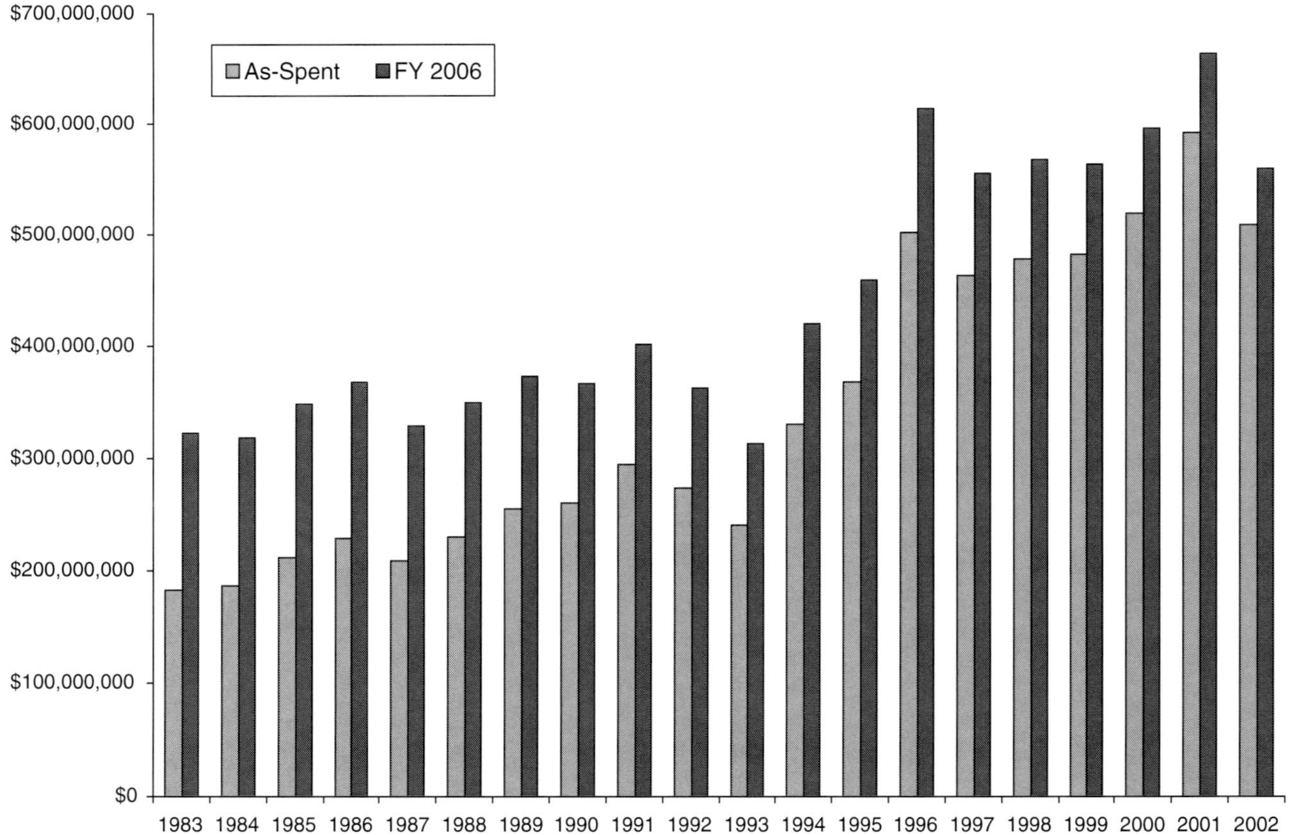

FIGURE 2.2 History of federal support for basic materials research, 1983-2002, excluding facilities construction and operations. In inflation-adjusted dollars, the federal investment in basic materials research has grown by more than 80 percent since 1983 but has remained essentially constant since 1996. This punctuated growth partly reflects the broadening of fields considered to be "materials research." SOURCE: National Science Foundation, *Science and Engineering Indicators 2006*.

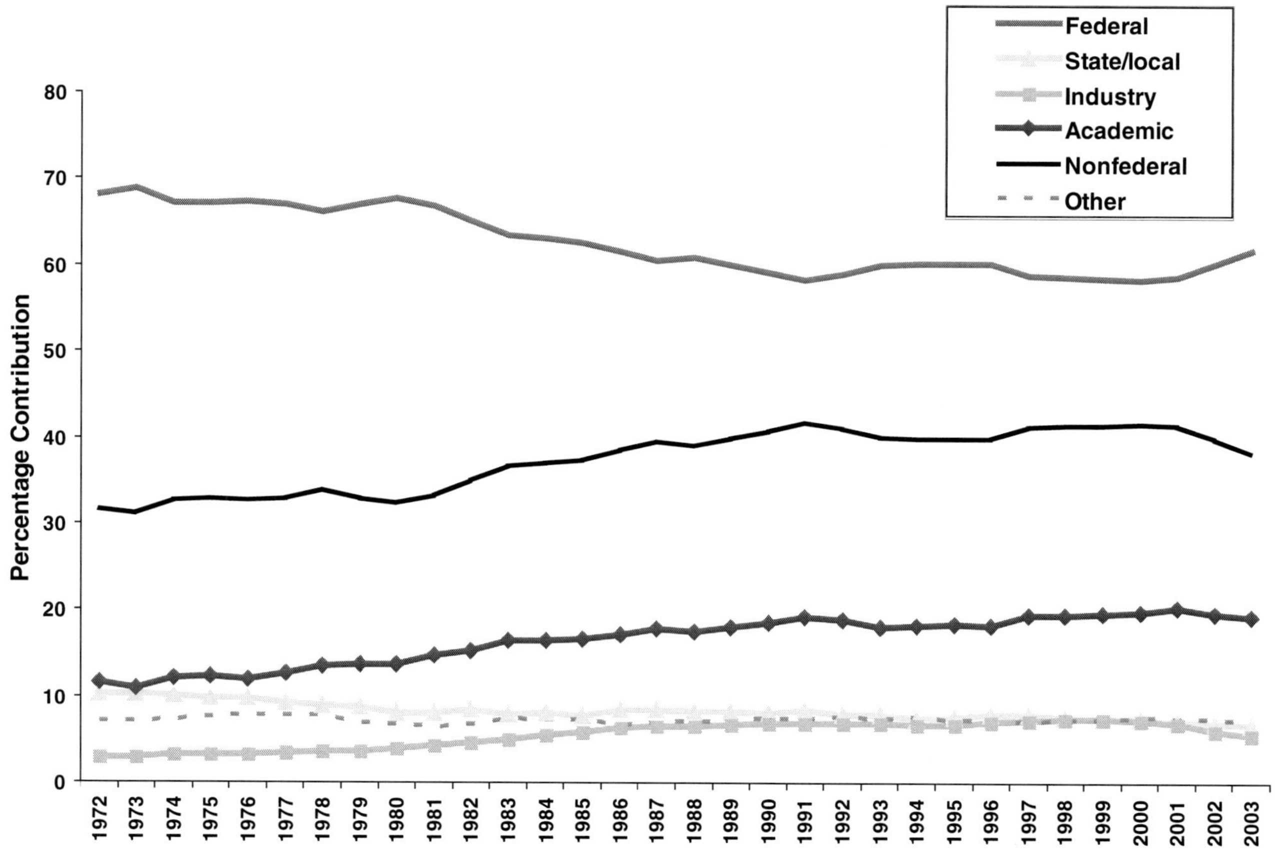

FIGURE 2.3 Sources of academic research and development funding for all research, 1972-2003. The funds provided for academic research and development by the industrial sector grew at a faster rate than funding from any other source during the 1972-2003 period. SOURCE: National Science Foundation, *Science and Engineering Indicators 2006.*

declined in both 2002 and 2003, the first time that such a decline had occurred in the past three decades. As a result, industry provided only 5% of academic research and development (R&D) funding in 2003, a substantial decline from its peak of 7% in 1999. Industrial support accounts for the smallest share of academic R&D funding, and support of academia has never been a major component of industry-funded R&D. In 1994, industry's contribution to academic R&D represented 1.5% of its total support of R&D, compared with 1.4% in 1990, 0.9% in 1980, and 0.7% in 1973. Between 1994 and 2004, this share declined from 1.5% to 1.1%.

NSF and the Division of Materials Research

MRSECs were created from the MRL program beginning in 1994, with all MRLs either terminated or converted to MRSECs by the end of 1996. Also by the end of 1996, many new centers were created, resulting in a total of 24 MRSECs. At the same time, the budget for MRL/MRSEC centers increased from approximately $29 million per year (as-spent dollars) in 1993 to $44.28 million per year in 1996. This represented a change of 124 percent in the number of centers but only a 53 percent increase in the total budget (see Figure 2.4). Clearly, MRSECs were "designed" to be smaller than MRLs, and some of the functions of the MRLs were eliminated. In most cases, the MRL-MRSEC transition trimmed staff in shared experimental facilities (SEFs) and decreased the rate and value of equipment purchases for such facilities. Since that time, the MRSEC as-spent budget first slowly increased and then essentially reached a plateau during the years 2003 to 2006 (now at $53.4 million per year).

An interesting comparison is between the "average budget" of an MRL in 1993 and the "average budget" of a MRSEC in 1996 and 2006. To make the comparison realistic, some method of taking inflation into account must be factored in. NSF has used an "OMB [Office of Management and Budget] inflation index"; a second option is the Consumer Price Index (CPI) for all consumers; and finally, there is a "university inflation index."[15] The first two are not identical, but perhaps close enough to follow NSF in the use of the OMB index (for example, from December 1994 to December 2005, the CPI increased by 31.5 percent, while the OMB index increased by 23.9 percent).

The committee estimated the university inflation index by determining the basic cost of a graduate student, taken as tuition, stipend, and overhead incurred

[15]The committee acknowledges that an inflation index for university research is not standard practice. However, informal discussions with deans of research programs revealed a growing interest in employing such a tool. For additional information on this topic, the committee refers readers to the more detailed discussion in the National Research Council report *Condensed-Matter and Materials Physics: The Science of the World Around Us*, Washington, D.C.: The National Academies Press, 2007.

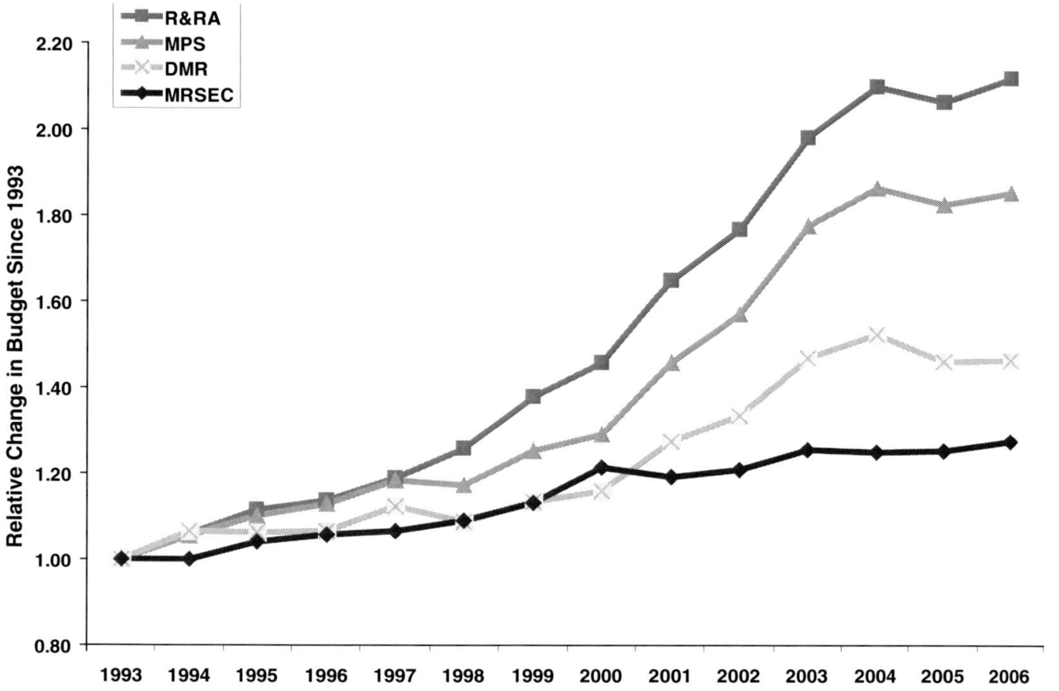

FIGURE 2.4 Cumulative relative change in as-spent budget for different parts of the NSF funding stream: the Materials Research Science and Engineering Centers (MRSEC) program, the Division of Materials Research (DMR), DMR's parent directorate of Mathematical and Physical Sciences (MPS), and the overall research and related-activities (R&RA) expenditures at NSF. Note that the MRSEC budget line did not formally start until 1994.

on the stipend. Not included in the index are health care, research equipment, and typical materials and supplies. Since "university inflation" as described above is not tracked by any agency, data were obtained from six institutions that have MRSECs. The sample included both private and state universities. For the period December 1994 to December 2005, the lowest growth index value was 52%, with the majority in the range between 70% and 82%. In any case, a safe average of 70% for university inflation is used for this period, acknowledging that the true average rate may be ±10% different from that value. It is also important to note that the rate is not uniform from university to university because each university faces a different set of circumstances.

In 1993, the 10 MRLs had an average budget of $2.9 million (as-spent). Using the OMB index, this adjusts to $3.65 million per year or, using the university inflation index, adjusts to $5.0 million per year. Table 2.3 shows the data for 1993 (MRLs only), 1996 (MRSECs fully established), and 2005.

TABLE 2.3 Annual Budget for an Average MRSEC-Type Center in Different Years Using Different Inflation Indexes

Year	Budget/Center (as-spent) (U.S. $)	Adjusted with OMB Index to FY 2005 (U.S. $)	Adjusted with University Index to FY 2005 (U.S. $)
1993	2,900,000	3,650,000	5,000,000
1996	1,850,000	2,200,000	2,750,000
2005	2,000,000	2,000,000	2,000,000

NOTE: OMB, Office of Management and Budget.

Given the decrease in spending power in the university environment, the average MRSEC can undertake only about 70 percent of the "effort" (as measured by financial investment) that it undertook in 1996, and only 40 percent of the effort that an MRL could undertake in 1993. A second way to express this decreased effort is to look at the total MRSEC budget from 1996 to 2006, which, when adjusted for university inflation, has decreased by 22 percent. Thus, a current MRSEC has fewer financial resources at its command than a previous MRL had. And so, are MRSECs necessarily accomplishing less in comparison? Because the scope of MRSEC activities is so different from that of MRLs and because the research has evolved, it is hard to draw a firm conclusion.

To put this in perspective, first compare these figures to the budgets of NSF and DMR, respectively. According to NSF data, the NSF budget for research and related activities (uncorrected for inflation) increased from $2.046 billion to $4.333 billion from 1993 to 2006 (or an increase of 112 percent, a number that is substantially above university inflation). The situation for DMR is dismal by comparison: from 1993 to 2006, the budget increased from $175.3 million to $242.9 million (or by 38 percent, somewhat more than the OMB inflation index but well below the university index).

Figures 2.5 through 2.7 give the details of the DMR trends. These cost comparisons do not correspond with the number of students reported as supported by NSF for the MRSECs. For example, data supplied by NSF suggest that the number of graduate students and postdoctoral (PD) associates supported in the MRSEC program has increased from 238 + 88 (PD) to 990 + 319 (PD), or an overall increase of 400 percent, although the "start-up time" of matriculating graduate students into the MRSEC program at the time of its inception causes significant distortion (see Figure 2.8). Clearly, the students counted are receiving partial support (much less than half). Clearly, the number of full-time-equivalent (FTE) students and PDs supported by DMR and the MRSEC program must have *decreased* over the past decade. This is exactly what the committee heard from numerous PIs in its visits to universities around the country. This angst has been met by ingenious ways

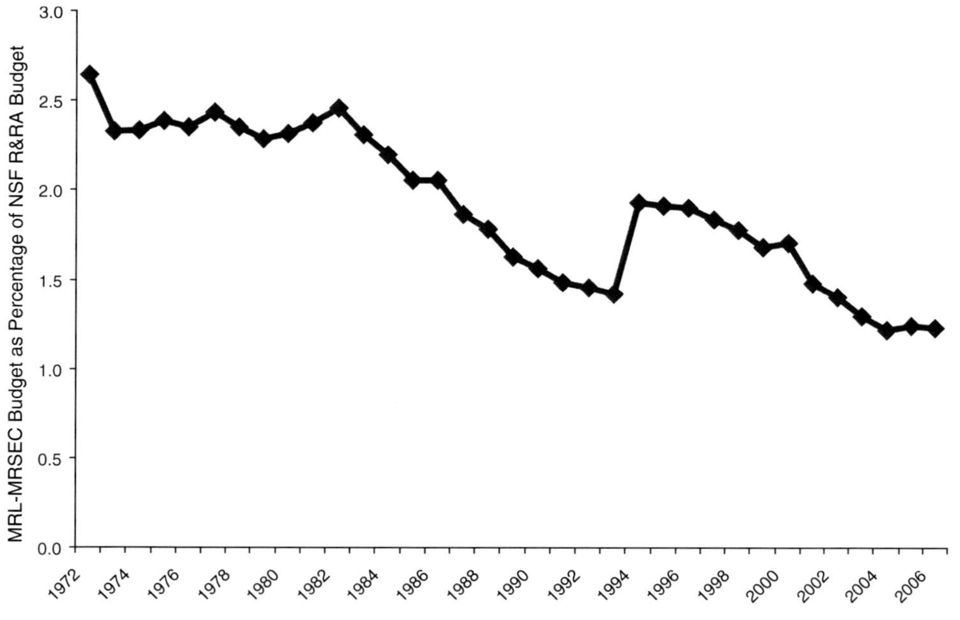

FIGURE 2.5 History of the percentage of National Science Foundation Research and Related Activities (R&RA) budget spent on the Materials Research Laboratory program (up through 1993) and the MRSEC program (starting in 1994), 1973-2006. SOURCE: Division of Materials Research, National Science Foundation.

of bringing multiple sources of funding to bear in order to advance the materials research field, thereby blurring the boundaries further between MRSEC- and non-MRSEC-supported research. These observations beg the question, however, of whether there is any direct linkage between MRSEC impact and partial support of students. The committee did not derive a quantitative metric, but it did come to believe that letting the escalating trend of engaging more and more students with less and less per capita resources constitutes a dilution of impact, not a continuous improvement in efficiency.

The data shown in Figure 2.8 are compiled from MRSEC annual reports. Those reports obviously include people who are partially supported by MRSEC and therefore also by other (unidentified) funds. If one wants to measure how many students the MRSEC influences, the currently reported data are more appropriate. Indeed, the true number is larger than that at institutions where the MRSEC runs extensive facilities, since many students supported on individual NSF grants as well as other types of support (DOE, state, and so on) use those facilities. If, however, one wants to focus on the overall MRSEC research effort, the FTE number would be more appropriate. To enable more consistent reporting over time, it might be

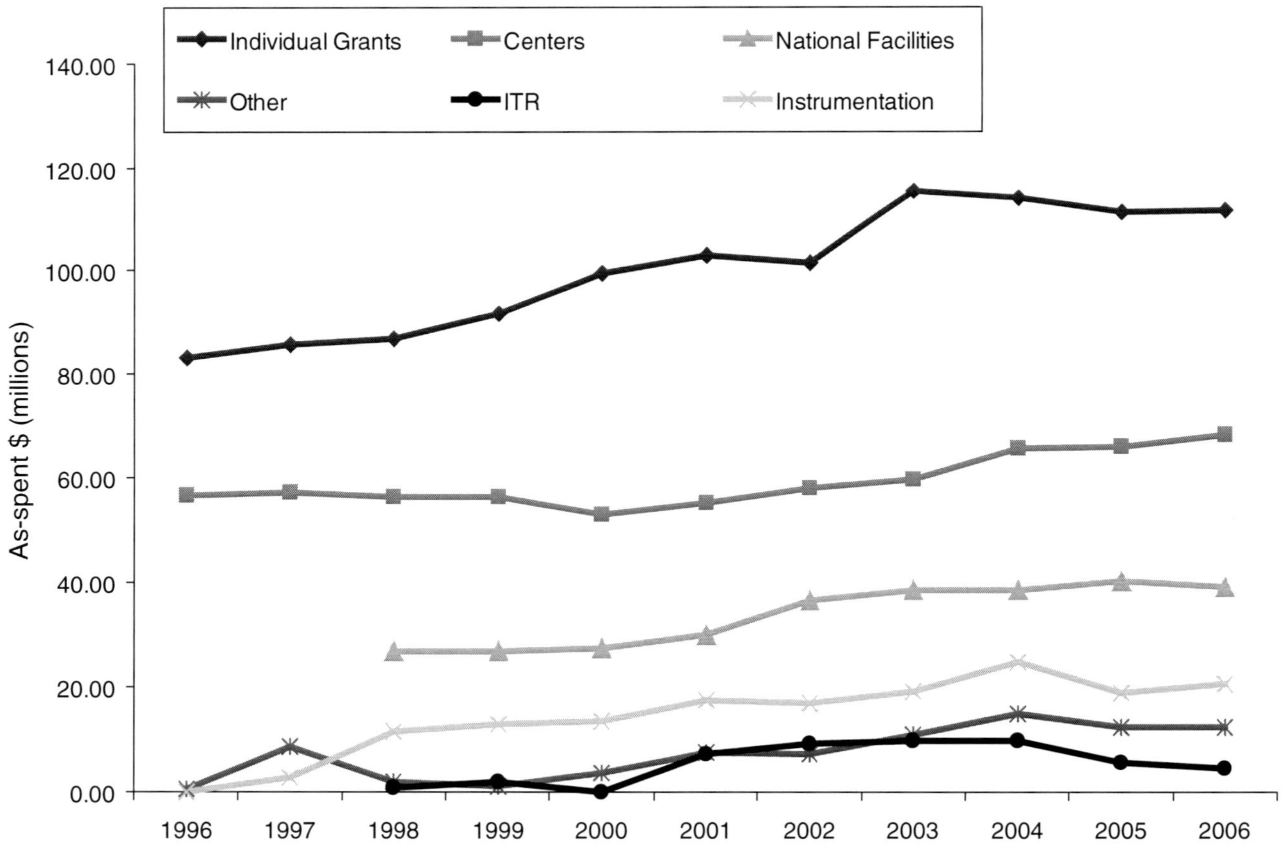

FIGURE 2.6 As-spent dollars for various programs and activities in the NSF's Division of Materials Research from 1996 to 2006. Centers include Materials Research Science and Engineering Centers, Partnerships for Research and Education in Materials, and some contribution to Nanoscale Science and Engineering Centers and Science and Technology Centers. The individual-investigator programs have increased by 34 percent in this period (but they have been decreasing slightly in the past 3 years), the centers by 20 percent, national user facilities by 45 percent (but the committee has data only for 8 of the 10 years), and instrumentation (Instrumentation for Materials Research program and Major Research Instrumentation, although the latter is non-DMR funds) by 42 percent. The MRSEC part of the centers program has increased in this period by 20.5 percent. ITR, Information Technology Research.

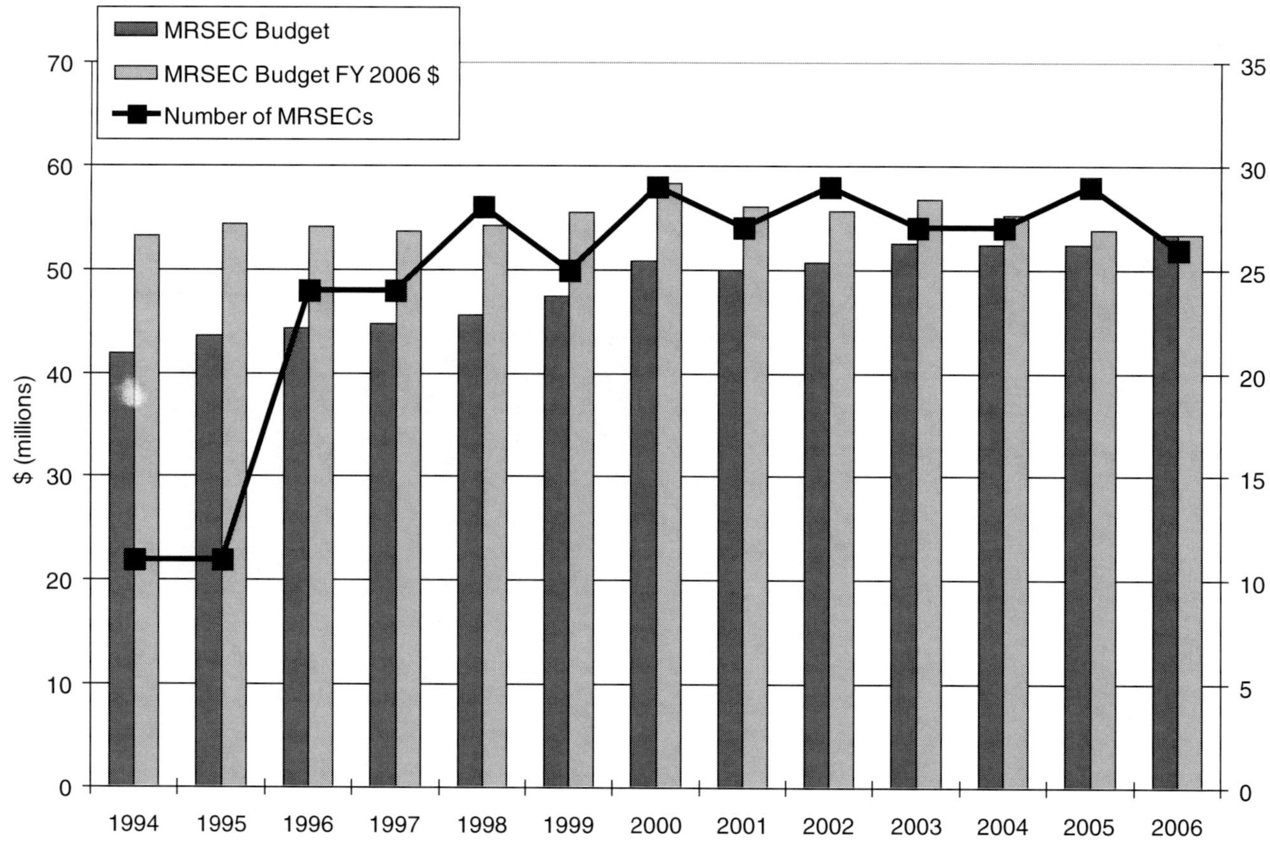

FIGURE 2.7 Annual budget, 1994-2006, for the Materials Research Laboratory/Materials Research Science and Engineering Centers program shown in as-spent dollars and in constant dollars as determined by the Office of Management and Budget inflation index; superposed (black line) is the number of MRSECs operating each year. Note that in 2006, three MRSECs were being phased out and were receiving partial funding in the phase-out period. While this plot suggests that MRSECs have been essentially flat-funded for the past 12 years, the estimated average university inflation index suggests a decline in spending power. SOURCE: Division of Materials Research, National Science Foundation.

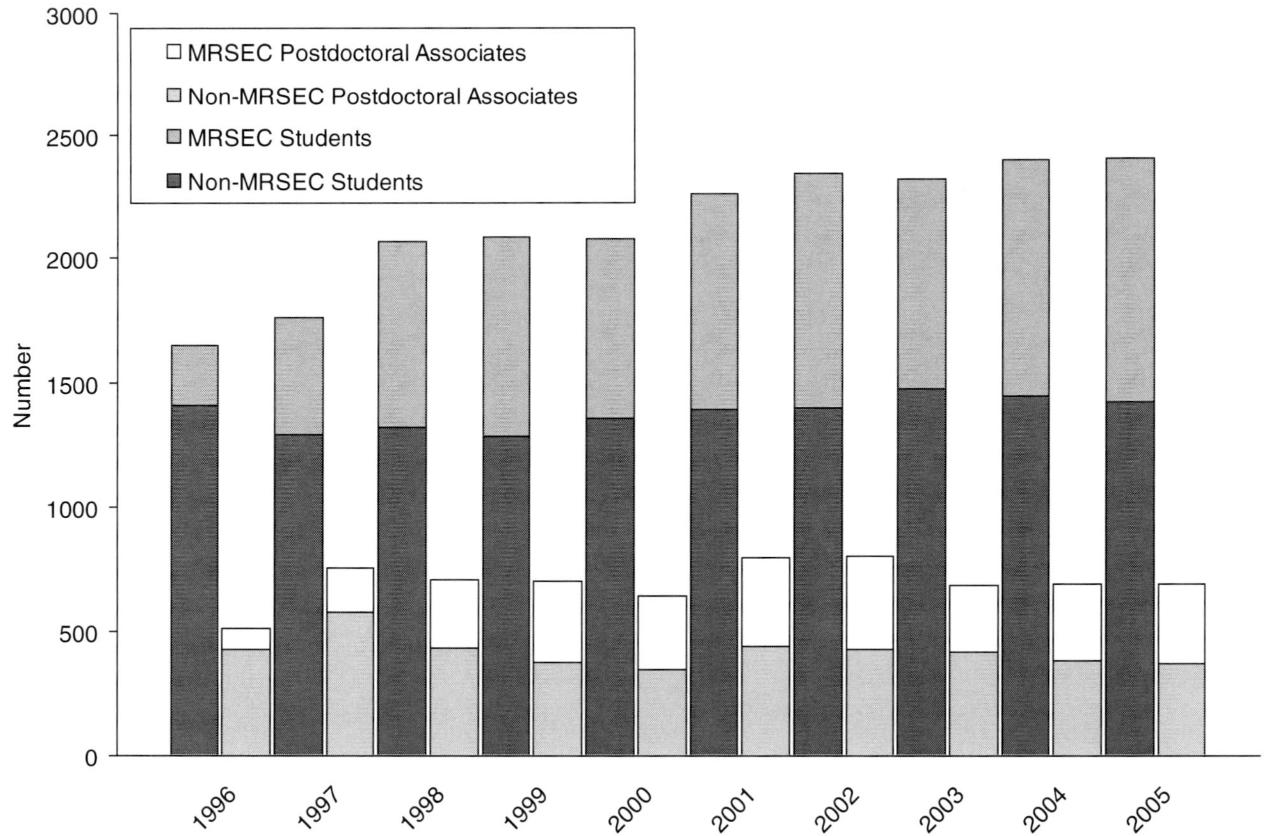

FIGURE 2.8 History of the participation of both students and postdoctoral associates in NSF's Division of Materials Research (DMR) programs overall and in the Materials Research Science and Engineering Centers (MRSEC) program specifically (as recorded by NSF's tabulation of annual reports from each MRSEC), 1996-2005. As explained in the text, the expanding number of students participating in MRSECs reflects non-full-time-equivalent reporting. SOURCE: Division of Materials Research, National Science Foundation.

useful for the MRSEC program directors to propose “full-time equivalent” units when centers report levels of participation in the program.

Single investigators at DMR have faced similar conditions. From 1996 to 2005, the median DMR single-investigator grant increased from $83,786 to $112,333 in as-spent dollars, an increase of 34%. During this time, the number of grants increased from 377 to a high of 561 and then decreased to 365 in order to increase the average size of the grants. While the size of the grants has increased in as-spent (34%) and even in OMB-inflated dollars (27%), it has not kept pace with university inflation (an average of 70%) and is much less than the overall increases in the NSF budget in as-spent dollars (more than 100%).

It is certainly possible that the materials community has not been making its case at NSF, and especially at OMB and in Congress. In comparison with other research fields, the community has not been able to articulate adequately the grand visions for the future and the potential benefits to the nation and society in general. Even within the materials field, activities (workshops, reports, and conferences) convened by the Office of Basic Energy Sciences at DOE have been much more successful in making the case for “use-inspired” research within the mission of that agency.

Program Evolution and Turnover

Of the 10 MRLs that existed in 1993, 8 are functioning MRSECs in 2006. These are the MRSECs at Brown University, the University of Chicago, Cornell University, Harvard University, the Massachusetts Institute of Technology, Northwestern University, the University of Massachusetts, and the University of Pennsylvania. Of these, all but the University of Massachusetts MRSECs are rooted in the IDL program (see Table 2.2 and Figure 2.1). Since 1996, when there were 24 MRSECs, 10 have been terminated and 13 started, leading to a total of 26 MRSECs in 2006 (not counting 3 that are receiving phase-out funds). Of the MRSECs added since 1994, a few have grown to be “large MRSECs,” with 3 or more IRGs (Princeton University; University of California, Santa Barbara; and Pennsylvania State University [PSU] in particular, although PSU did host an MRL that was terminated in 1980). Most of the rest are smaller MRSECs with 1 or 2 IRGs. Turnover in the program indicates that the peer-review process managed by NSF does have some impact. The committee was not in a position to second-guess any particular award decisions; numerous committees of visitors to NSF’s Division of Materials Research have affirmed the integrity of the process.

In the last MRSEC competition, only 2 new MRSECs were added to the program out of more than 100 preproposals, with 3 existing centers being phased out. The committee notes that the low success rate represents a substantial amount of effort. Excessive as this may seem, the 100 preproposals submitted to NSF indicate that the effort is still worthwhile and that MRSEC program is highly sought.

Current MRSEC Budgets

In 2006, the MRSEC budget at NSF was $53.48 million per year. There are 26 MRSECs and 3 in phase-out funding, so the average MRSEC budget is close to $2 million per year (but the actual range is $1.0 million to $3.8 million per year, not counting Partnership for Research and Education in Materials [PREM] funding). As seen in Table 2.4, the MRSEC budget is divided into 6 categories: IRGs (63%), Seeds (10%), Facilities (11%), Educational Outreach (10%), Industrial Outreach (2%), and Administration (4%). As with the individual MRSEC budgets, there is considerable variability from center to center in these categories, especially in the last three (see Figure 2.9). Individual MRSECs also leverage these funds through institutional commitments, user fees in shared experimental facilities, and/or industrial and state support.[16]

It is interesting that the decrease in support (at the university inflation rate) for both the MRSECs and single investigators has put an even larger strain on maintaining forefront facilities. Since SEFs rely on a large user base for user fees, and since many of the users are supported on single-investigator grants with shrinking materials and supplies budgets, the facilities system is being squeezed from both sides. Neither MRSEC nor single-investigator research in materials can be competitive worldwide (or even carried out) without the capabilities present in SEFs. In fact, this was one of two principal aims of the IDL program when it was first established (the other was to promote interdisciplinary research). Successful industrial collaboration relies, more often than not, on good instrumentation and facilities on the academic side of the collaboration to enable the exploratory research sought by the industrial partner. The importance of shared experimental facilities and the availability of capital and operating funds cannot be overestimated.[17]

NSF plans for the future of the materials center program must address this issue or the materials program will soon be noncompetitive on the international level. The National Academies report *Experiments in International Benchmarking of US Research Fields* states that "there continues to be concern among top university researchers that facilities and equipment for materials research in several foreign

---

[16]The committee heard testimony during its site visits that research groups proposing MRSECs were backed by as high as a 30% cost-share from the host institutions alone.

[17]See also, National Academy of Sciences, National Academy of Engineering, and Institute of Medicine, *Advanced Research Instrumentation and Facilities*, Washington, D.C.: The National Academies Press, 2006, especially Chapters 3 and 4.

TABLE 2.4 Breakdown of Average Annual Budget for a Materials Research Science and Engineering Center in 2006

Category	Average MRSEC Spending per Year (U.S. $)
Interdisciplinary Research Groups	1,260,000
Seeds	200,000
Facilities	220,000
Educational outreach	200,000
Industrial outreach	40,000
Administration	80,000
Total	2,000,000

SOURCE: B. Keimer, Max Planck Institute for Solid-State Research, private communication.

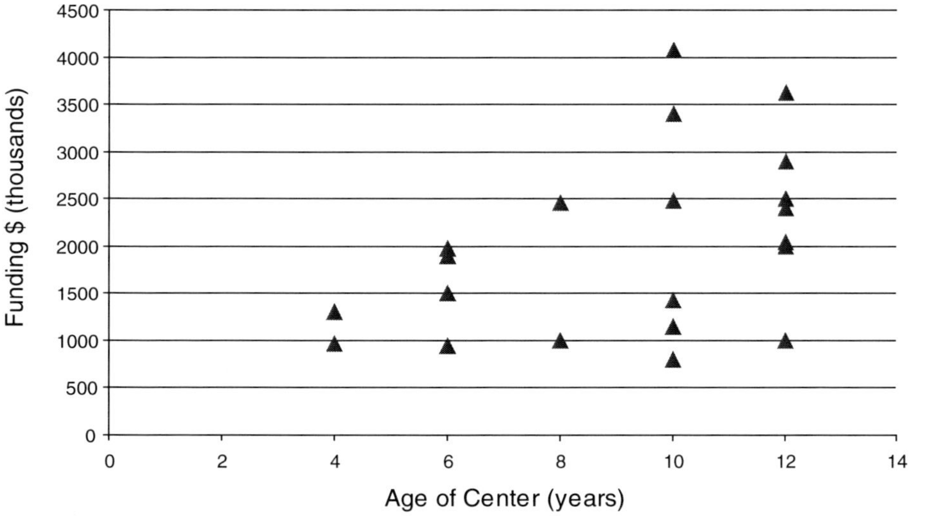

FIGURE 2.9 Materials Research Science and Engineering Center (MRSEC) annual budgets as reported in their FY 2005 annual reports versus the age of the materials center at the host university. Of the 15 centers in the 10- and 12-year bins, 9 centers received funding beginning with the Interdisciplinary Laboratory or Materials Research Laboratory program. SOURCE: Compiled by the committee from collected annual reports (FY 2005) from the MRSECs.

universities now outclass those at most universities in the United States."[18] Indeed, during a site visit, one individual observed that resources for instrumentation and facilities in the United States were so poor that the MRSEC-added SEF funding merely slowed the local rate of decay as compared with other U.S. facilities, thereby maintaining relative leadership.

[18]National Academy of Sciences, National Academy of Engineering, and Institute of Medicine, *Experiments in International Benchmarking of US Research Fields*, Washington, D.C.: National Academy Press, 2000, pp. 2-26.

INTERNATIONAL CONTEXT

Research centers are a common element of national materials research programs in many countries. For instance, the United Kingdom, Germany, Switzerland, France, Japan, and China all have systems of research centers as part of their public investments in materials research. Thus, the U.S. MRSEC program is not globally unique. In this section, the committee briefly examines the international landscape of materials research to put the U.S. MRSECs into a global context.

Other reports have made significant strides in characterizing the U.S. materials research enterprise in comparison with foreign programs. For instance, the National Research Council report *Globalization of Materials R&D: Time for a National Strategy*[19] presents a framework for developing a strategic approach to national research efforts in an increasingly connected world. While this committee examined that report and similar ones, it made no effort to repeat the analysis. Rather, this committee comments here on the role that materials research centers play in several other countries.

The Deutsche Forschungsgemeinschaft (DFG, German Research Foundation) is responsible for the management of a new federal program called the "Excellence Initiative" for strengthening research at German universities. The German federal and state governments have provided a total of €1.9 billion for 5 years to boost research performance at Germany's top universities; a further 5 years are envisaged. The money will support approximately 30 clusters of excellence (about €6.5 million per year each) and approximately 40 graduate schools (about €1 million per year each) and will fund structural measures to enhance international competitiveness. The first round of evaluations is now finished. A total of 292 draft proposals for graduate schools and centers of excellence were reviewed in different panels. As a result of this first evaluation step, 41 initiatives for clusters of excellence and 38 initiatives for graduate schools were invited to submit full proposals, among these are 3 clusters of excellence and 3 graduate schools in the field of "Condensed Matter Sciences." After a total of 88 proposals for the three funding lines were evaluated and discussed by international review panels and the Joint Commission of the German Science Council and the DFG, the Excellence Initiative Grants Committee awarded funding to 18 graduate schools, 17 clusters of excellence, and 3 institutional strategies. The decisions were announced in Bonn by the Federal Minister of Education and Research, Dr. Annette Schavan, as well as the Ministers of Science and Research, Professor Peter Frankenberg (Baden-Württemberg), and Professor

[19]National Research Council, *Globalization of Materials R&D: Time for a National Strategy*, Washington, D.C.: The National Academies Press, 2005.

Jürgen Zöllner (Rhineland-Palatinate). For this first round, about €175 million have been approved per year to fund initiatives at 22 universities.[20]

The National Research Council report *Midsize Facilities: The Infrastructure for Materials Research* made the following observations about activities in Japan and elsewhere in Europe:[21]

> Some important features are revealed by considering how these same issues are approached in other countries. Japan has the extremely impressive National Institute for Materials Science (NIMS) in Tsukuba, with about a thousand researchers and a remarkable array of equipment (e.g., over 35 advanced TEMs, including two high-voltage, high-resolution microscopes that no longer exist in the United States after the decommissioning of the NCEM microscope in 2004). The Japanese facilities, like those in the major national universities, reside largely in the groups of individual investigators rather than being multiuser operations.
>
> In the United Kingdom, excellent facilities are found at the elite institutions (e.g., Oxford and Cambridge Universities). These are continually upgraded (e.g., Oxford already has an aberration-corrected TEM) and are well supported with technical and scientific staff. However, there tend to be fewer users from outside those institutions.
>
> Many of the midsize facilities in France are supported by the Centre National de la Récherche Scientifique (CNRS). Thus, many of the scientists are permanent government employees themselves. . . . The smaller European countries with on the order of 10 million population each (e.g., Belgium, the Netherlands, Sweden) tend to have a few highly funded, well-supported centers which are national resources and are extensively used by many colleagues on a national and international level. Examples include the high-resolution electron microscope (HREM) laboratory at the Middelheim Campus, University of Antwerp, Belgium; the Dutch National Center for HREM in Delft, Netherlands; the Swedish National Center for HREM at the Lund Institute of Technology; Interuniversity MicroElectronics Center (IMEC) in Leuven, Belgium; and Materials Analysis at Chalmers (MACH) at Chalmers University of Technology, Göteborg, Sweden. The model of these smaller countries is one to note especially. These are stable, well-funded facilities that serve a large number of users. They are successful because of a combination of recognized need, enthusiastic collaboration, and continued oversight from the government and scientific community. It is also undoubtedly advantageous that these countries are geographically small, so that national facilities are never more than a few hours' drive away.

The 1998 report of the National Research Council's Committee on Science, Engineering, and Public Policy entitled *International Benchmarking of US Materials*

---

[20]For the full list of awards, see the press release at http://www.dfg.de/en/news/press_releases/2006/press_release_2006_54.html (last viewed October 14, 2006).

[21]National Research Council, *Midsize Facilities: The Infrastructure for Materials Research*, Washington, D.C.: The National Academies Press, 2006, p. 27.

Science and Engineering Research[22] presented an assessment of the U.S. position in materials science and engineering (MSE) research in the near and long term, based on current trends in the United States and abroad. The report concluded that the United States is among the world leaders in all subfields of MSE (as defined in the report). It does warn, however, that the United States should expect an erosion of leadership as Europe and Japan increase their support of MSE. The 2005 report *Globalization of Materials R&D*[23] also shows that while the U.S. share of global R&D has remained steady since the 1990s, its lead in MSE is weakening and being tested by the European and Asian regions.

[22]National Academy of Sciences, National Academy of Engineering, and Institute of Medicine, *International Benchmarking of US Materials Science and Engineering Research*, Washington, D.C.: National Academy Press, 1998, pp. 73-77.

[23]National Research Council, *Globalization of Materials R&D: Time for a National Strategy*, Washington, D.C.: The National Academies Press, 2005, pp. 30, 160-169.

3

Assessment of Research and Facilities Impact

In carrying out its impact assessment task, the committee first analyzed the issues broadly in several different categories: research, education and outreach, collaboration with other sectors, and other areas. After the following introductory remarks, an analysis of the impact of Materials Research Science and Engineering Centers' (MRSECs') research is presented in this chapter.

INTRODUCTION

The Materials Research Science and Engineering Centers program (MRSEC program) was established by the National Science Foundation (NSF) in its Division of Materials Research (DMR) in 1994. As described in Chapter 2, the MRSEC program was born out of the decision to transform the Materials Research Laboratory (MRL) and Materials Research Group (MRG) programs to the current structure. The goal of this new initiative was to provide focused support for complex interdisciplinary materials research and education at the university level. To receive a MRSEC award, an institution must:

> [demonstrate] outstanding research quality and intellectual breadth, provide support for research infrastructure and flexibility in responding to new opportunities, and strongly emphasize the integration of research and education. These centers foster active collaboration between universities and other sectors, including industry, and they constitute a national network of university-based centers in materials research. MRSECs address problems of a scope or complexity requiring the

advantages of scale and interdisciplinary interaction provided by a campus-based research center.[1]

Awards granted under the program provide support for a 6-year period; during the last 2 years of this period there is an external review under recompetition requirements in the program's language. After the original competition in 1994, additional competitions occurred in 1996, 1998, 2000, 2002, and 2005. At the inception of the program in 1994, 30 full proposals were submitted, and 11 awards were given to 9 universities. Owing to the phase difference in the transition between programs, 13 new MRSEC awards were granted 2 years later.

The program currently funds about 29 MRSECs (26 active MRSECs and 3 on phase-out funding), which split a total of about $51 million with a range of $1.0 million to $5.0 million per institution per year, as shown in Figure 3.1. The awards are fully competed every 6 years but are staggered on the basis of the date of first award. An institution that does not receive continued MRSEC funding after recompetition is provided with phase-out support. The total number of MRSECs funded since the program's inception in 1994 is as follows:

Year:	1994	1996	1998	2000	2002	2004
No.	11	24	25	27	27	27

A MRSEC provides a forum for researchers to come together and to share thoughts and ideas. Researchers participate because they realize the great advantages of working in an interdisciplinary team with exciting colleagues. The long-term nature of MRSEC support is welcomed because it allows researchers to pursue high-risk but potentially transformative ideas. Those ideas may lead to a new research direction for the MRSEC or may gain funding from other sources. MRSECs also provide a context for pursuing fundamental research that may not have immediately obvious payoffs but that is critical to future discovery. Students working within a MRSEC have a unique opportunity to learn from multiple mentors and to gain experience with techniques and ideas outside their own immediate field. Speaking to the committee, Harvard University MRSEC Director David Weitz emphasized these points by stating, "The most important products of the MRSEC are ideas (science, start-ups, etc.) and well-trained people."

Evaluating MRSEC research is a daunting task. The committee considered several strategies, realizing that the MRSEC program contributes to the NSF mission in multiple ways even though "short-term research results" are usually considered the primary objective (see Box 3.1).

[1]National Science Foundation, *Program Solicitation for Materials Research Science and Engineering Centers*, NSF 04-580, Washington, D.C., 2004.

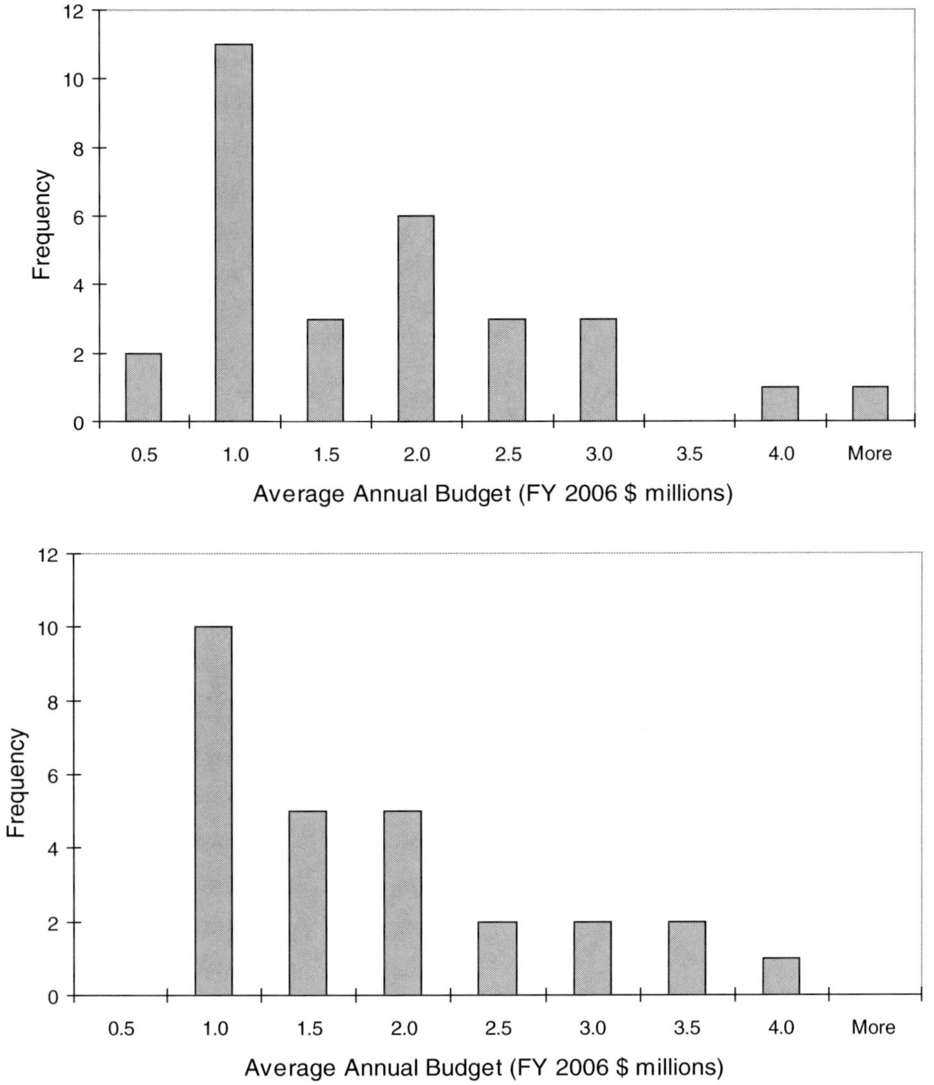

FIGURE 3.1 Distribution of annual Materials Research Science and Engineering Center budgets in the 1990s (top) and in the current decade (bottom). The average (median) center budget in the late 1990s was $1.5 million ($1.2 million in inflation-adjusted dollars) and is now $1.6 million ($1.3 million); the width of the distribution has narrowed slightly. The axis label for each bar on the histograms indicates the upper edge of the range of values assigned to that bin. The data for the 1990s include several centers that started and stopped during the last years of that decade. SOURCE: Division of Materials Research, National Science Foundation.

BOX 3.1
Qualitative Tests of MRSEC Impact

A necessary exercise in assessing the impact of the Materials Research Science and Engineering Centers program (MRSEC program) is to look at the top-rated programs in materials research in the United States and to compare them with a list of institutions that have Materials Research Science and Engineering Centers (MRSECs). Because it is impossible to determine the causality between the existence of a MRSEC at an institution and the quality of that institution's materials science and engineering program, the committee conducted a very cursory examination of several surveys, as discussed below. At best, the surveys exhibit a correlation.

Surveys by *U.S. News and World Report*

According to an annual survey conducted by *U.S. News and World Report,* the top five schools for undergraduate materials science and engineering schools in 2006 were the following:[1]

1. Massachusetts Institute of Technology (MIT)
2. University of Illinois at Urbana-Champaign
3. { Northwestern University
   University of California at Berkeley
   University of Michigan at Ann Arbor }

Note that the last three schools tied for third place. Four of these schools have had MRSECs or Materials Research Laboratories (MRLs).

In 2007, according to *U.S. News and World Report,*[2] the top four chemistry Ph.D. programs were at the California Institute of Technology, MIT, Stanford University, and the University of California at Berkeley. Three of these schools have MRSECs. Likewise, the top graduate physics programs were at MIT, Stanford University, and the California Institute of Technology. All three schools have MRSECs, although the Stanford MRSEC does not interact broadly with the physics department. Drilling down to the level of condensed-matter and materials physics, both of the top schools (Cornell University and Harvard University) have MRSECs closely connected with the physics departments; the University of Illinois at Urbana-Champaign has a Materials Research Laboratory that is now supported by the Department of Energy (DOE).

National Research Council Ph.D. Program Rankings

The National Research Council conducts a decadal survey of graduate programs. The most recent rankings are from 1995, with the next edition expected in 2008. When completed, the new rankings will nicely

[1] *U.S. News and World Report,* 2006. Available in limited form online at http://colleges.usnews.rankingsandreviews.com/usnews/2006/edu/college/rankings/rankindex_brief.php.
[2] *U.S. News and World Report,* 2007. Available in limited form online at http://colleges.usnews.rankingsandreviews.com/usnews/2007/edu/college/rankings/rankindex_brief.php.

The committee realized that a comparison ("control") group would need to be defined in order to make substantive comparisons. For instance, it would be insufficient to observe, "Research conducted through the MRSEC program generally includes significant collaboration." Rather, the committee sought to determine

bracket the first decade of the MRSEC program. In 1995, the top 10 graduate programs in materials science were as follows:[3]

1. MIT
2. Northwestern University
3. Cornell University
4. University of California at Berkeley
5. University of Illinois at Urbana-Champaign
6. Stanford University
7. University of Massachusetts at Amherst
8. University of California at Santa Barbara
9. Pennsylvania State University
10. University of Pennsylvania

Of these 10 schools, all but the University of California at Berkeley have a formal materials centers program dating back to the Interdisciplinary Laboratories of the 1960s and the Materials Research Laboratories of the 1970s and 1980s. The University of California at Berkeley has strong connections with the neighboring DOE Lawrence Berkeley National Laboratory as well as with Stanford's Center on Polymer Interfaces and Macromolecular Assemblies.

National Doctoral Program Survey

In 2000, the National Association of Graduate and Professional Students published the results of a survey of over 32,000 participants.[4] The survey ranked graduate programs on the basis of participants' perception of the overall implementation of recommended best practices (admittedly nebulous). At the top of the list of materials science programs ranked by recommended practices were the following schools:

1. MIT
2. University of Massachusetts at Amherst
3. Johns Hopkins University
4. Pennsylvania State University
5. Stanford University
6. University of Delaware
7. University of California at Berkeley
8. University of Minnesota

Of these schools, all but the University of California at Berkeley and the University of Delaware have (had) MRSECs.

[3]National Research Council, *Research Doctorate Programs in the United States: Continuity and Change*, Washington, D.C.: National Academy Press, 1995.
[4]Available online at http://cresmet.asu.edu/nagps/about/index.php.

if the rate or nature of collaboration in the MRSEC program is different from the rate or nature of collaboration outside the program. A natural control group might therefore be the body of research enabled by the individual-investigator awards made through NSF's Division of Materials Research. If a positive measurement were obtained, one might then ask the importance of "group-based research" to the nation's research enterprise.

This approach was complicated by the fact that research papers published in peer-reviewed journals do not, in general, uniquely attribute the research results to a single mechanism of support. No researcher finds his or her support entirely from a MRSEC, and both MRSEC and other contributions are influenced by participation in the MRSEC. Even at the level of an individual researcher's activities, it is categorically impossible to separate out the uniquely "MRSEC-enabled" research. Site visits confirmed these impressions (see Box 3.2).

This caveat should be borne in mind when interpreting these analyses. Despite this intrinsic limitation, however, the committee designed and carried out several exercises, examining the activities enabled by the MRSEC program using several different techniques to "separate out" the MRSEC contributions and to construct "control" groups for comparison.

Finally, the committee emphasizes that its goal was not to evaluate the MRSEC program specifically, nor to recommend the continuation or termination of the program, but rather to describe and characterize its impact. Ideally, the committee would like to have answered a pointed question: If one had the opportunity to reinvest the annual budget of the MRSEC program purely on the grounds of its research impact, are there compelling examples of "what could not have happened otherwise?" Unfortunately, the inability to clearly separate what is "because of MRSEC" from "what is not" made it impossible to answer this question.

Moreover, any research, even by an individual researcher associated with a MRSEC, is a combination of activities supported both inside and outside the MRSEC. Thus, even if MRSECs have played a unique role in the research enterprise, such as in enabling the formulation of research projects that could not otherwise have been envisioned, there is no easy way to provide substantiation. Although the committee was unable to identify MRSEC-enabled research in "blind taste tests," it successfully assessed the overall research quality in comparison with the research enabled by other mechanisms and elsewhere around the world. The basic questions to be answered are whether the research enabled by the MRSEC program is distinctive, if it is worthwhile and of high quality, and, finally, whether it is a good investment.

Many studies have tried to assess the quality of research programs in terms of objective criteria such as the citation numbers. A previous evaluation of the MRL program by the MITRE Corporation for the NSF concluded that there were no discernable trends in the quantity of publications or their citations when comparing MRLs with similarly funded programs. The committee's exploration of the citation index produced similar conclusions. The committee found that identifying a set of comparable institutions was difficult, that the data would not be easy to obtain, and that the results would at best indicate the *average* output of the MRSECs and

BOX 3.2
Site Visits

The MRSEC Impact Assessment Committee conducted more than a dozen site visits at institutions that either have a MRSEC or a similar center-based research structure, that were contemplating a MRSEC application, or which had had a MRSEC that closed (see the section entitled "Site Visits" in Appendix F, "Data-Gathering Tools," for a list of the institutions visited). These visits prompted candid conversations with researchers that provided valuable anecdotal information and firsthand impressions. The committee found this feedback very useful in its assessment of the MRSEC program. Below are some excerpts from conversations with faculty and staff.

- From its site visits, the committee heard comments that, *"Centers can only succeed if they help us integrate between disciplines,"* and that *"Without impetus from outside [the university], it is hard to initiate a center, despite whatever latent good will and intentions there are."* When asked to differentiate the MRSEC-style research center from departmental centers, some said, *"Centers are intellectual foci of effort. They are at a larger scale than just one department with some of its faculty. You need to cut across more fields of research to really attack new problems and push forward; you need more than one or two departments."* When asked about their university's perspective on centers, many university administration officials commented that they view centers with federal funding as having a higher degree of validity because they have received some external commitment and recognition.
- When asked about canceling the MRSEC program, one university official opined, *"I don't think that the campus and state would take the initiative to invent such a center without the external incentive unless the affected topics were related to human health and wellness. Also, this campus is isolated geographically from the industrial community,"* and would not be as able to engage industries in relationships pertaining to physical science projects. Others echoed these thoughts, saying that such centers are one of the only mechanisms, externally funded, that cut across disciplinary boundaries and the stovepipes of academic departments. Others commented that single-investigator awards are typically only 3 years, and the longevity of a center grant enables much more creativity, flexibility, and even security in trying out research ideas. A final comment suggested that the National Science Foundation fulfills a key national goal by providing support for basic research that is not directly connected to product commercialization (as opposed to state and local industry programs), and since centers are a key mechanism for supporting the basic-science enterprise, they should be continued.

their comparison group. In most areas of endeavor, it is not the average that leads to remarkable advances but, rather, remarkable discoveries that are large fluctuations from the norm.

While it was difficult to separate research uniquely enabled by the MRSEC program from research that was made possible by other means, the committee was clearer about causation. For instance, many of the more recently established NSF Nanoscale Science and Engineering Centers (NSECs) are located at institutions that have MRSECs; of the 10 active NSEC awards, 3 are at institutions without active MRSECs, and at least 1 more is in a research area very different from the corresponding MRSEC. Do MRSECs enhance the probability for NSEC awards? Or does experience in the MRSEC competition simply add to an institution's competitive edge? One could wonder about the potential for a "chicken-and-egg" problem at a strong institution that was awarded a MRSEC: which came first, the strong campus research effort or the center? In the committee's judgment, the competitive selection process for MRSEC awards puts the burden on the pre-existing strength of the institutional research effort. While a MRSEC may enhance an institution's materials research programs, it simply cannot bring them into being.

ANALYSIS OF SELECTED CONTRIBUTIONS FROM MATERIALS RESEARCH

The committee identified a seemingly promising exercise that ended up rather inconclusively. Based on personal judgment and discussion with colleagues, the committee constructed a list of selected important materials discoveries and inventions from 1960 to 2000 (see Figure 3.2 for the subfields in which these discoveries occurred). Because the list was subjective, the committee chose not to publish it here. The committee then identified where the research leading to each discovery had been done and, in particular, whether it had originated in the MRSEC program or its predecessor MRL program. The list contained very few items that occurred in the past decade and thus the discoveries on the list significantly predated the MRSEC program per se. While this list is admittedly subjective and does not purport to be definitive, it revealed that the number of discoveries attributable to U.S. universities is rather limited. Given the generally recognized quality of U.S. universities in materials research, it is surprising that only 5 of these 27 discoveries are attributable to U.S. universities. Three were attributable to MRSECs. The committee does not want to overstate the implications of this ad hoc analysis, but it at least suggests that MRSEC research is an important part of a U.S. university materials research portfolio (see Box 3.3).

The majority of the discoveries were undertaken by individuals or small groups of about two investigators. Many of the discoveries originated in the predominantly industrial research laboratories in the United States, which reflected the time period (1960-2000) considered. Many of these labs (AT&T, IBM, Xerox, GE, Exxon, and so on) have been greatly reduced or eliminated, raising important questions about whether MRSECs can compensate for these losses. One should note, however, the parallel between a scientific breakthrough and car racing—the car and the driver get all the credit for a win, but in truth a much larger team is needed to enable a victory.

A small fraction of these breakthroughs took place in universities with MRSECs or MRLs (see Figure 3.3). Although this may appear somewhat discouraging regarding the impact of MRSECs as primary sources for innovative materials, funding

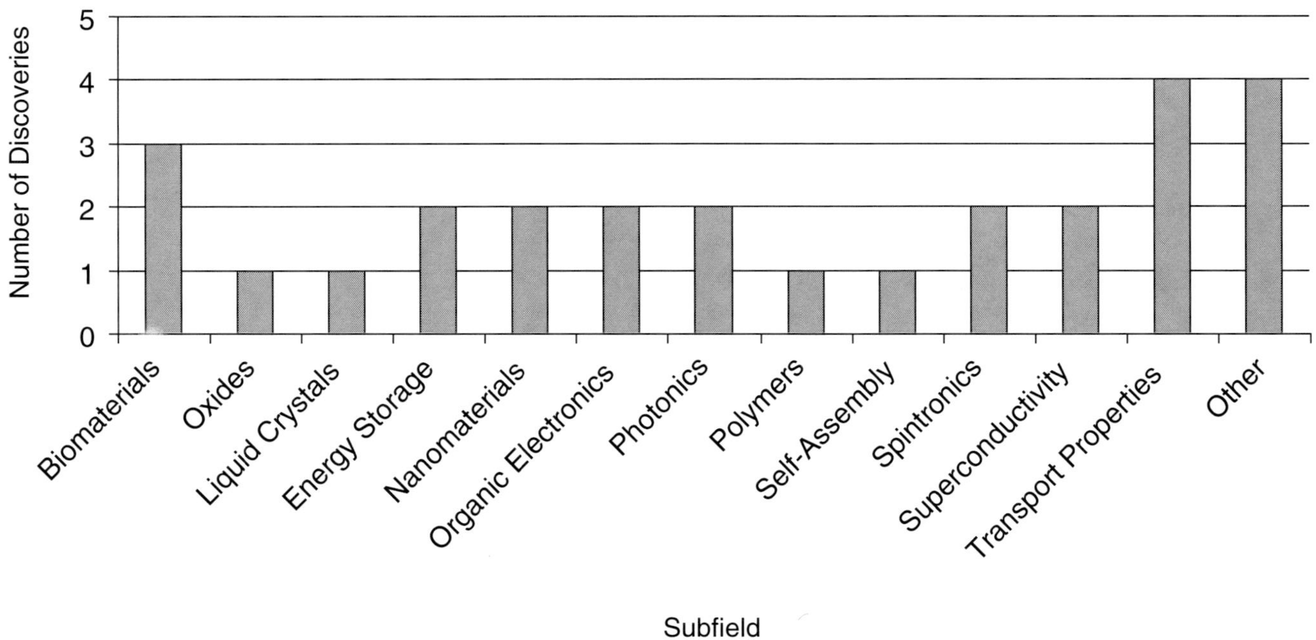

FIGURE 3.2 Distribution of selected major materials research discoveries, by materials research subfield. A handful of the 27 discoveries identified by the committee did not fit into these categories and are labeled "Other."

BOX 3.3
The MRSEC at the California Institute of Technology

One example of a successful MRSEC is at the California Institute of Technology (Caltech). A thumbnail sketch of Caltech's Center for the Science and Engineering of Materials provides a view of the typical organization and breadth of activities of a MRSEC. The Caltech MRSEC focuses on several interdisciplinary areas.

The research program is organized into four interdisciplinary research groups (IRGs) and two seed projects. The IRGs are Biomolecular Materials, Ferroelectric Thin Films, Mesophotonics, and Bulk Metallic Glasses. The work in Biomolecular Materials explores the control of self-organization that can be achieved in polymers of absolutely defined comonomer sequence—genetically engineered artificial proteins—and the control of spatial arrangement and size that can be templated using surfactant nanostructures. The Ferroelectric Thin Films group aims to enable ultrahigh displacement microactuators based on high-strain ferroelectrics. The project on Mesophotonic Materials is motivated by advances in the synthesis and theoretical understanding of materials designed to manipulate light on scales at and below the wavelength of light, in order to move into the revolutionary domain of devices on scales of tens of nanometers.

The program on Bulk Metallic Glasses, which has been particularly effective in its industrial interactions, investigates the processing, microstructure, and mechanical behavior of bulk metallic glasses (BMGs) and their composites. The researchers are investigating the basic science and engineering that will enable new strategies to produce BMGs in which a crystalline phase is introduced to resist shear localization, creating a BMG composite with enhanced material properties. Some of this work has already reached the stage of commercial application in such products as cellular telephone cases and other electronic device packages. Future efforts, in conjunction with a number of industrial partners, involve applications in a wide variety of commercial, biomedical, and military applications.

levels must be considered. The total budget of the MRSEC program is about $50 million per year compared with about $2 billion per year spent for (basic and applied) materials research by the U.S. government, and the approximately $4 billion per year spent worldwide by governments on materials research. The contributions seem to be larger than might be expected simply from the funding ratio. The fraction of MRSEC dollars to total materials dollars is 0.05/4, or 1.25 percent. No statistical analysis of these fragmentary observations is possible; however, it is possible to say that there are discoveries of the highest significance occurring within the MRSEC program, as gauged by this subjective survey.

As can be seen later in this section, there is almost an orthogonality between the types of institutions responsible for "major discoveries" and "top cited papers," the former originating in industrial laboratories and the latter in universities. The committee suggests that many of the more recent fundamental breakthroughs occur in academe, often with MRSEC-funded facilities, whereas materials discoveries more closely linked to commercial products were more naturally done in industrial settings. The trend may reflect the passing of the torch from the formerly powerful industrial labs to universities.

The field of organic/polymeric conductors was supported from its inception

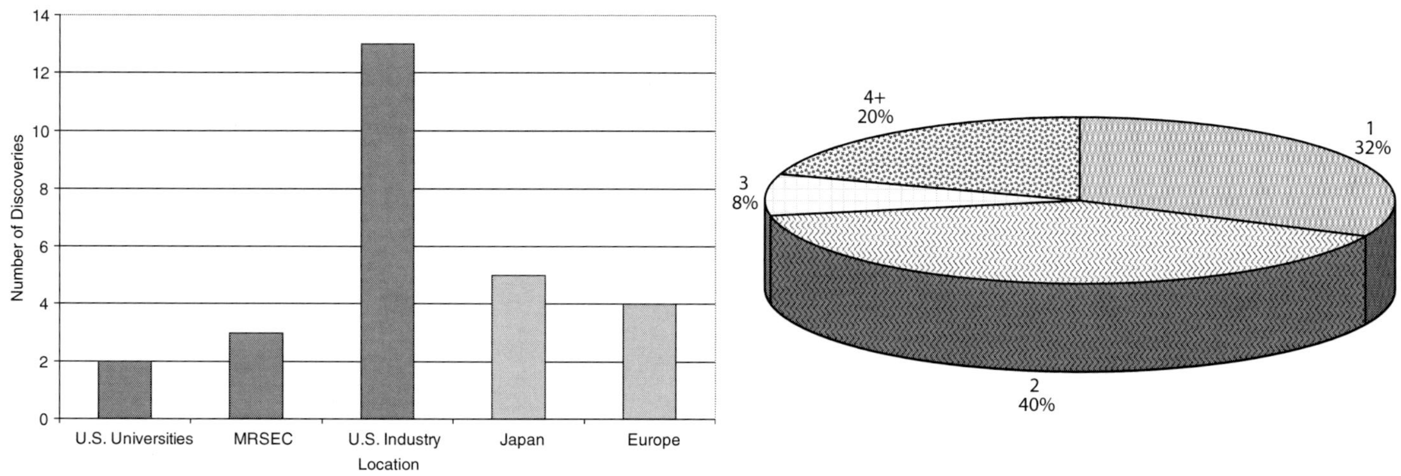

FIGURE 3.3 Characteristics of the most significant discoveries in materials research. Left: Where each breakthrough occurred. Note the predominance of U.S. industry. Right: Number of senior researchers involved in each discovery. Note that most discoveries were made by teams of only one or two senior investigators.

BOX 3.4
Progress on Conducting Polymers

Alan Heeger began his university career at the University of Pennsylvania (Penn) in 1961, just one year after Penn was established as one of the initial three materials Interdisciplinary Laboratories (IDLs) (see Table 2.2 in this report). His initial research was in metal-insulator transitions, particularly in one-dimensional systems. This early work was the genesis of Heeger's work with Alan MacDiarmid leading to the development of plastic electronics. Alan Heeger was director of the Penn Materials Research Laboratory (MRL) when this work accelerated with the synthesis of polyacetylene published in 1977 by Shirakawa, MacDiarmid, Loius, Chiang, and Heeger (see Figure 3.4.1). Financial support from the MRL, along with major support from the Office of Naval Research, enabled these seminal discoveries that led to the award of the 2000 Nobel Prize in chemistry to Shirakawa, MacDiarmid, and Heeger. In 1982, Alan Heeger moved to the University of California at Santa Barbara, where soon after he successfully proposed a new Materials Research Group on conducting polymers. This became the nucleus for the founding of the University of California at Santa Barbara MRSEC in 1993.

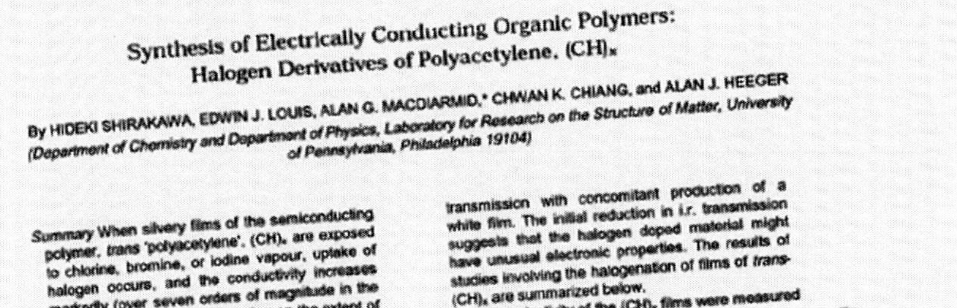

J.C.S. CHEM. COMM., 1977, 578-580

Synthesis of Electrically Conducting Organic Polymers: Halogen Derivatives of Polyacetylene, $(CH)_x$

By HIDEKI SHIRAKAWA, EDWIN J. LOUIS, ALAN G. MACDIARMID,* CHWAN K. CHIANG, and ALAN J. HEEGER
(Department of Chemistry and Department of Physics, Laboratory for Research on the Structure of Matter, University of Pennsylvania, Philadelphia 19104)

Summary When silvery films of the semiconducting polymer, *trans* 'polyacetylene', $(CH)_x$ are exposed to chlorine, bromine, or iodine vapour, uptake of halogen occurs, and the conductivity increases markedly (over seven orders of magnitude in the case of iodine) to give, depending on the extent of ... silvery or silvery-black films, som...

transmission with concomitant production of a white film. The initial reduction in i.r. transmission suggests that the halogen doped material might have unusual electronic properties. The results of studies involving the halogenation of films of *trans*-$(CH)_x$ are summarized below.

The condu...vity of the $(CH)_x$ films were measured

The original 1977 *ChemComm* paper

FIGURE 3.4.1 Top: Alan Heeger. Bottom: Image of the 1977 Nobel Prize–winning paper.

in the MRL program (see Box 3.4). This highlight of the MRL/MRSEC program is on the list of selected major materials discoveries (see Box 3.5 for other examples of MRSEC research). Early contributions to the fundamental physics of quasi-one-dimensional conductors as new materials were produced in the MRLs, in U.S. universities, and in European and Japanese laboratories. A breakthrough in

BOX 3.5
Examples of MRSEC Research

Magnetic Tubules: Cellular Tracks Follow the Field at Pennsylvania State University MRSEC

Motor proteins deliver intracellular cargo to specific locations inside cells. These so-called kinesin motors take 8 nanometer (nm) steps along intracellular highways, called microtubules, 25 nm wide. This transport machinery can be reassembled outside the cell and used to transport nanoscale cargo for separations, sensors, assembly, and other biomechanical devices. However, to fully harness these biological motors outside the cell, a means is needed both to attach cargo and to lay down the tracks at the desired locations and orientations. Materials Research Science and Engineering Center (MRSEC) researchers at Pennsylvania State University are using magnetic fields to control the placement and transport of microtubules. In the reverse of a mobile engine on a stationary railroad track, the biomotor track (the microtubules) is actually mobile while the motors (kinesins) are bound upside-down to the surface, ready to push the microtubule along like a person body surfing at a rock concert. Magnetic nanoparticles of $CoFe_2O_4$ are attached to the microtubules as magnetic "handles." By adjusting the ambient magnetic field, the microtubules can be reoriented, allowing them to be transported in any desired direction. Even weak magnets can direct the biomotor-driven transport of thousands of microtubules at once. Magnetically labeled microtubules also provide a new tool for investigating the role of microtubules and motors in cellular processes such as cell division, axonal transport, and flagellar motility.

High-Performance Transparent Inorganic-Organic Hybrid Thin-Film n-Type Transistors at Northwestern University MRSEC

Thin-film transistors, already indispensable in a number of portable electronics, would benefit from optical transparency and compatibility with flexible, lightweight plastics. Transistors with these qualities would be a major advance if they could be fabricated by a scalable, large-area process. Researchers at the Northwestern University MRSEC have adopted a hybrid approach in developing "invisible" thin-film transistors that heterogeneously integrate a transparent, inorganic semiconductor with a large carrier mobility and a nanoscopic, organic gate dielectric.

Bacterial Nanoreactors at University of Southern Mississippi MRSEC

The nanometer-scale polyhedral protein compartments (carboxysomes) found in many bacteria harbor an enzyme (RubisCO) that converts carbon dioxide to sugars, which in turn are used by the cell to synthesize other biomolecules. One of the carboxysome shell proteins (e-carbonic anhydrase) was found to catalyze the dehydration of bicarbonate and to direct the resulting CO_2 toward the inside of the carboxysome, where it is efficiently metabolized by RubisCO. It is currently thought that the orientation of the carbonic anhydrase in the protein shell constitutes a chemical diode that makes the carboxysome shell directionally permeable to CO_2 and allows it to function analogous to polymer film-immobilized catalysts. Work is currently under way to understand the self-assembly of carboxysome protein components, which ultimately may guide efforts to synthesize selectively permeable protein-based films for potential pharmaceutical or manufacturing applications.

the synthesis and characterization of polyacetylene turned the field around. This collaborative research was conducted at an MRL and involved strong interactions between the physics and chemistry departments at Pennsylvania State University and a group in Japan; it led to new materials, conducting polymers, and new concepts: solitons, fractional charge, spin-charge separation. Further developments by several groups led to new technology, especially for organic optical displays; the formation of several companies in the United States and overseas; and the awarding of the 2000 Nobel Prize in chemistry to Alan J. Heeger, Alan G. MacDiarmid, and Hideki Shirakawa.

The committee experimented with a related exercise by trying to identify the founding papers in several topical areas of research. For instance, in the field of magnesium diboride research, the breakthrough paper is clearly the 2001 *Nature* article by Nagamatsu and colleagues (cited in the list below). Employing a scientific publication citation analysis tool (Scopus), the committee identified the top 20 most highly cited (subsequent) papers that cited the founding paper. Using institutional affiliation and some knowledge of the MRSEC IRG membership, the committee examined the role of MRSECs in this new set of "soon afterward" papers. The results were largely inconclusive, but interesting nonetheless.

- *MgB_2:* J. Nagamatsu, N. Nakagawa, T. Muranaka, Y. Zenitani, and J. Akimitsu, "Superconductivity at 39 K in Magnesium Diboride," *Nature* 410 (6824): 63-64, 2001. Cited 1,804 times.
  —Of the top 20 most highly cited articles and reviews, none was from an institution with a MRSEC.
- *Spintronics:* Y. Ohno, D.K. Young, B. Beschoten, F. Matsukura, H. Ohno, and D.D. Awschalom, "Electrical Spin Injection in a Ferromagnetic Semiconductor Heterostructure," *Nature* 402 (6763): 790-792, 1999. Cited 867 times.
  —Of the top 20 most highly cited articles, 1 was from an institution with a MRSEC.
- *Magnetic semiconductors:* T. Dietl and H. Ohno, "Zener Model Description of Ferromagnetism in Zinc-Blende Magnetic Semiconductors," *Science* 287 (5455): 1019-1022, 2000. Cited 1,284 times.
  —Of the top 20 most highly cited articles, 2 were from institutions with MRSECs.
- *Novel oxides:* P. Schiffer, A.P. Ramirez, W. Bao, and S.-W. Cheong, "Low Temperature Magnetoresistance and the Magnetic Phase Diagram of La1-xCaxMnO3," *Physical Review Letters* 75 (18): 3336-3339, 1995. Cited 1,061 times.
  —Of the top 20 most highly cited articles, 5 came from institutions with MRSECs.
- *Stripes in high-temperature superconductivity:* J.M. Tranquada, B.J. Sternlieb, J.D. Axe, Y. Nakamura, and S. Uchida, "Evidence for Stripe Correlations of Spins and Holes in Copper Oxide Superconductors," *Nature* 375 (6532): 561-563, 1995. Cited 1,266 times.

—Of the top 20 most highly cited articles, 6 were from institutions with MRSECs.

PUBLICATION CITATION ANALYSES

A study of highly cited papers was conducted to obtain a more objective overview of current and past research directions and impact in the materials community. A potential metric for examining the overall impact of published research results is to consider publication citations—the number of times subsequent papers refer to an earlier work; one variation on this approach was already discussed. A number of assumptions need to be made to imbue this technique with credible utility.[2]

The committee notes an important limitation of this analysis in advance: because the MRSEC program is only about one decade old, any research publications authored under its auspices are still relatively nascent in the field. That is, truly eminent articles generally take 10 to 15 years to demonstrate their impact on the field.[3] Thus, the committee's efforts to assess the research impact of the MRSEC program through a study of its publication citations is a bit premature. In its defense, the committee chose to compare the MRSEC program to just the past decade of materials research papers in order to include the same systematic error in the reference case. In order to avoid replicating the MITRE report described earlier and in order to keep its task tractable and focused on the MRSEC program, the committee chose not to analyze the full legacy of the IDLs and MRLs that led to the MRSECs. However, in so doing, the committee's analysis could be interpreted to simply conclude that the MRSEC program is too young for impact assessment.

In a larger sense, the committee also sought to investigate some of the "urban myths" surrounding the MRSEC program. The committee made the most progress in addressing the question of whether MRSEC-enabled research results were empirically distinguishable in character and/or quality from other research. Box 3.6 shows some characteristic publications from the MRSECs.

Top 100 Most Highly Cited Papers in Materials Research

Using a scientific journal publication citation tool (Scopus), the top 100 most highly cited papers in materials research since about 1996 were identified. A break-

[2]See, for instance, David Adam, "Citation Analysis: The Counting House," *Nature* 415: 726-729 (2002) for a discussion of the intrinsic limitations of these techniques.

[3]Using the Essential Science Indicators tool provided for public use online by the ISI Web of Knowledge, the 100 most highly cited papers of all time in the field of materials science were queried. Of these 100, more than 40 percent were published before 2000.

BOX 3.6
Selected Research Results Enabled by the MRSEC Program

Despite the difficulty in empirically separating MRSEC-enabled research from research supported by other mechanisms, the MRSEC Impact Assessment Committee was able to examine a set of research papers self-reported by the MRSECs. Many of these had high citation indices; a very small subset are listed here as examples.

- J. Needleman, "A Continuum Model for Void Nucleation by Inclusion Debonding," *Journal of Applied Mechanics-Transactions of the ASME* 54 (3): 525-531 (1987). (Brown University MRL/MRSEC, 461 citations)
- J. Park, A.N. Pasupathy, J.I. Goldsmith, C. Chang, Y. Yaish, J.R. Petta, M. Rinkoski, J.P. Sethna, H.D. Abruña, P.L. McEuen, and D.C. Ralph, "Coulomb Blockade and the Kondo Effect in Single-Atom Transistors," *Nature* 417 (6890): 722-725 (2002). (Cornell MRSEC, 337 citations)
- C.S. Chen, M. Mrksich, S. Huang, G.M. Whitesides, and D.E. Ingber, "Geometric Control of Cell Life and Death," *Science* 276: 1425 (1997). (Harvard MRSEC, 910 citations)
- B.M. Discher, Y.-Y. Won, D.S. Ege, J.C.-M. Lee, F.S. Bates, D.E. Discher, and D.A. Hammer, "Polymersomes: Vesicles Made from Diblock Copolymers," *Science* 284: 1143 (1999). (Minnesota MRSEC, 310 citations)
- D.Y. Zhao, J.L. Feng, Q.S. Huo, N. Melosh, G.H. Fredrickson, B.F. Chmelka, and G.D. Stucky, "Triblock Copolymer Syntheses of Mesoporous Silica with Periodic 50 to 300 Angstrom Pores," *Science* 279 (5350): 548-552 (1998). (Santa Barbara MRSEC, 1,489 citations)
- A. Thess, R. Lee, P. Nikolaev, H.J. Dai, P. Petit, J. Robert, C.H. Xu, Y.H. Lee, S.G. Kim, A.G. Rinzler, D.T. Colbert, G.E. Scuseria, D. Tomanek, J.E. Fischer, and R.E. Smalley, "Crystalline Ropes of Metallic Carbon Nanotubes," *Science* 273 (5274): 483-487 (1996). (Pennsylvania State University MRSEC, 1,898 citations)
- Z.A. Peng and X. Peng, "Synthesis of High Quality Cadmium Chalcogenides Semiconductor Nanocrystals Using CdO as Precursor," *Journal of the American Chemical Society* 123: 168 (2001). (Oklahoma/Alaska MRSEC, 333 citations)

down of these hundred papers in terms of subfields is shown as the light-colored bars in Figure 3.4. The affiliation information from the citations was examined to determine several characteristics for each paper: nationality, national laboratory/industry/university origins, and MRL/MRSEC connection. The MRL/MRSEC connection was evaluated by determining whether the institution had an operating MRSEC at the time of publication.[4] The results are plotted in Figure 3.5. It is

[4]The committee notes that the assumption employed here is a weak point in the exercise. Materials research papers arising from institutions with MRSECs do not necessarily come from MRSEC-enabled research. However, the committee could not find a better alternative. Even cross-checking the principal investigators with MRSEC faculty lists would not work because faculty conduct research under many different auspices and with many different funding sources.

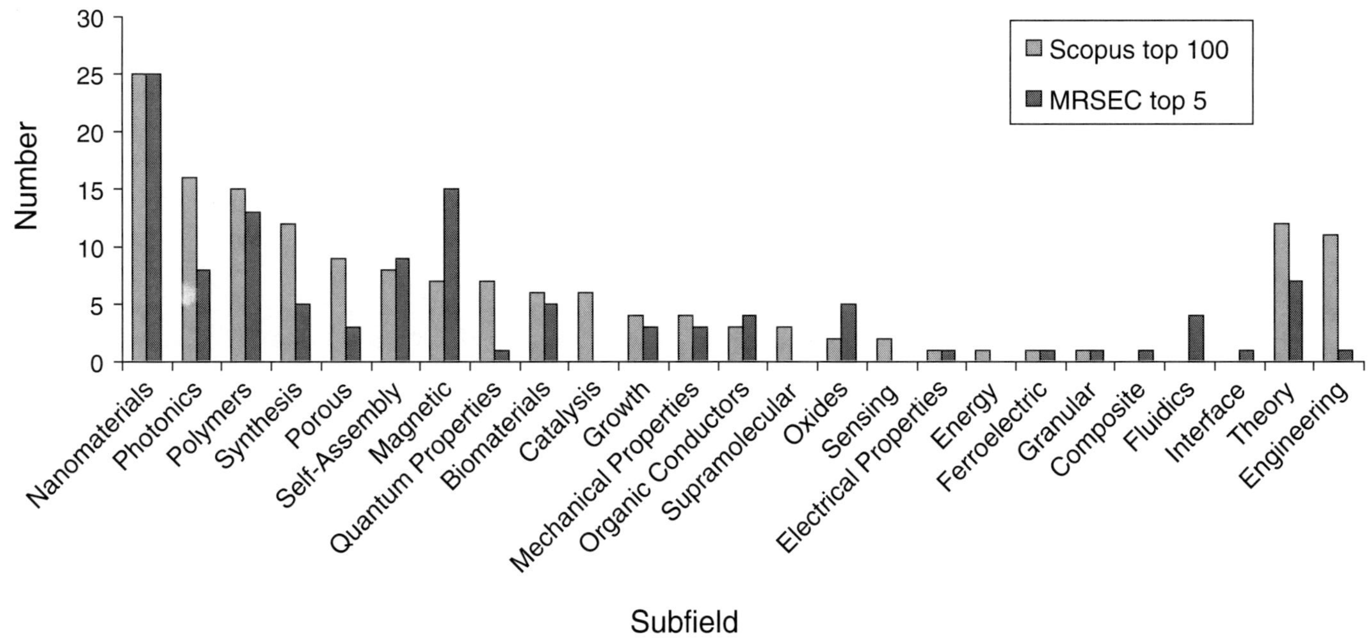

FIGURE 3.4 Grouping by subfield of the top 100 most-cited materials research papers (light bars). For comparison, the number of highly cited papers by subfield, for the set of "top 5" papers reported by the Materials Research Science and Engineering Centers, is included as the second bar (dark) in each category.

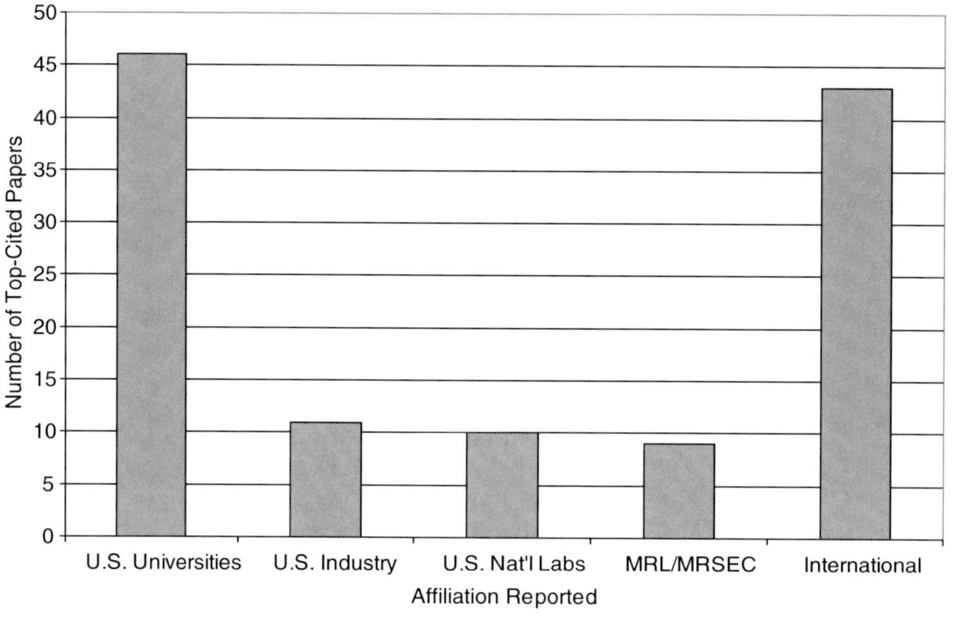

FIGURE 3.5 Analysis of the affiliations listed by the authors of the overall top 100 most highly cited papers in materials research, 1996-2006. The total number of entries exceeds 100 because the committee could not clearly assign the origins for several papers. NOTE: MRL, Materials Research Laboratories; MRSEC, Materials Research Science and Engineering Centers.

noteworthy that institutions with an MRL/MRSEC program accounted for about 10 percent of the most highly cited papers worldwide and about 20 percent of those from the United States. These are considerably larger percentages than might be indicated by the relative funding levels mentioned earlier. This is the first indication that institutions with MRSECs are among the leaders in materials research. However, the task of directly associating research quality with the MRSEC program is complex, since it cannot be distinguished whether the best institutions are likely to succeed in the MRSEC competition or whether the MRSECs play a dominant role in the materials effort at these institutions. It is probable that both effects are present. Separating institutional publication impact caused by the MRSEC program as opposed to simple correlation with a MRSEC (or even as part of the reason for winning a MRSEC) is hard. The difficulty in assigning credit for these highly cited papers is that most authors report support from many sources.

Portfolio of MRSEC Research Activities

At the committee's request, many of the current MRSECs provided a list of their 5 most cited papers over the past 10 years.[5] The distribution of MRSEC "top papers" by subfield was compared with the distribution of "top papers" for the entire materials field, thereby indirectly testing whether the MRSEC research portfolio matched the overall global materials research portfolio. A close examination of Figure 3.4 shows the result: the number of "top 100 papers" per subfield is plotted alongside the number of "top MRSEC" papers per subfield. The comparison indicates a strong correspondence between high-impact research done in the MRSEC program and the interests of the materials community as a whole.

The committee also independently examined the current MRSEC research portfolio. Although the MRSEC program is programmatically contained within the NSF Mathematical and Physical Sciences Directorate's Division of Materials Research (i.e., separate from the NSF Engineering Directorate), the intended scope of the MRSEC program includes materials engineering. The list of research topics studied by the current suite of MRSECs has very limited intersection with the engineering side of materials (see Appendix C for a list of the current IRG research topics).

Average MRSEC Citation Impact

To gauge the "average impact" of each MRSEC, the average citations for the top 5 papers from each MRSEC were computed (see Figure 3.6). Each entry represents the average citation count for one MRSEC. The average number of citations per MRSEC ranged from 37 to 994 per highly cited paper. Note that for the 100 most cited papers in materials research from 1996 to 2006, the average number of citations per paper was 892. Thus, the "best" materials research papers are better (in terms of citations) than the "best" MRSEC papers. Although it would not be reasonable to expect the best papers to arise exclusively from MRSECs, it is noteworthy that some of the best MRSEC papers do rate among the best papers overall. As shown in Table 3.1, the average citation rate for MRSEC-related papers is about 13. This average number weights over many broad valleys in addition to high peaks; for instance, younger MRSECs with less-established research programs in new areas tend to have lower citation rates and therefore pull the average down. However, the average citation count per paper for all materials science papers

[5]The committee used this "top 5" set in several key exercises; although self-reported by the MRSECs, it represented a list of more than 100 research results that MRSECs indicated playing a role in. The committee had no other reliable means of obtaining a set of MRSEC-enabled research papers beyond asking the MRSECs themselves.

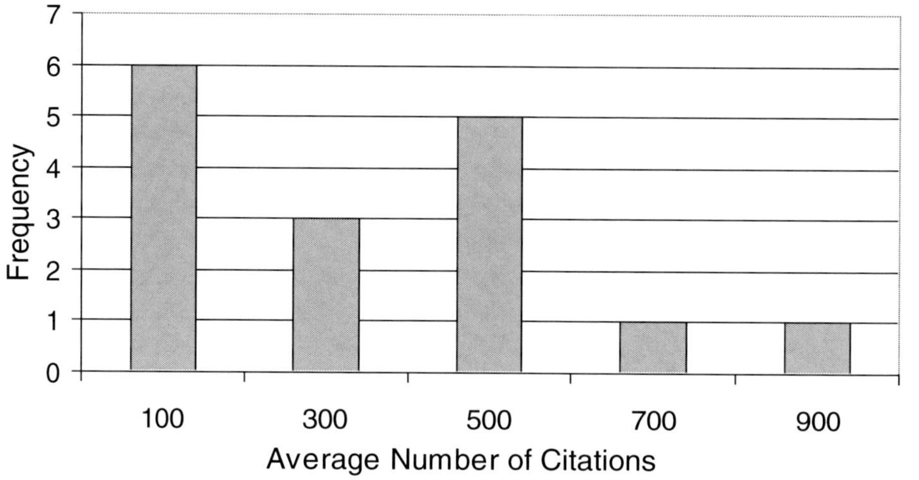

FIGURE 3.6 Distribution of average number of citations for the set of "top 5" papers reported by each currently active Materials Research Science and Engineering Center, 1994-2006. The lowest bin extends down to zero. The average of this histogram is about 350. The top 100 most-cited papers had an average of 892 citations per paper, with a median of 775.

TABLE 3.1 Impact of MRSECs Compared with the German System of the Max Planck Institutes

Research Institute	Publications 1995-1996 (No.)	Citations 1995-2006 (No.)	Citations per Publication (No.)
MRSECs	483	6,269	13.0
MPI-FKF	6,309	65,279	10.3
MPI-MF	3,316	31,349	9.5
MPI-POLY	3,455	48,162	13.9
MPI-MSP	1,781	19,427	10.9

NOTE: MRSECs, Materials Research Science and Engineering Centers; MPI-FKF, Max Planck Institute for Solid State Research, Stuttgart; MPI-MF, Max Planck Institute for Metals Research, Stuttgart; MPI-POLY, Max Planck Institute for Polymer Research, Mainz; MPI-MSP, Max Planck Institute for Microstructure Physics, Halle.

(368,111) is 4.48.[6] The top MRSEC papers thus are much better than the average materials research papers.

The committee compares the top MRSEC papers to the average papers for two reasons. First, it supports the observation that MRSECs have contributed several important research advances to the field. Second, the committee found it difficult to extract independently average values for the set of all MRSEC papers—mainly

[6]Similarly for physics, the average is 7.22 citations per for 834,162 papers; for chemistry the average is 8.24 citations per paper for 1,028,375 papers, according to Scopus.

because MRSEC papers are not well labeled in the literature; the only known data set is the one provided by MRSECs themselves. The committee found that the average MRSEC paper performed similar to the average materials paper.

Comparison of Citation Impact for Max Planck Institutes and MRSECs

The committee contacted institutions in other countries to discuss techniques for assessing the quality of their research activities, particularly those activities that are center-based. While there was no consensus as to the value of different procedures, many of the institutions contacted took what they considered was the "easy way out" and used publication-citation indices. Bernhard Keimer at the Max Planck Institute in Stuttgart kindly offered the services of his library staff to compare citations for several Max Planck Institutes (MPIs) and the MRSEC program. The search picked out papers in which a MRSEC was explicitly listed in an author's address field; thus, many papers were missed if the author had a different home department as his or her address. Nonetheless, the comparison, shown in Table 3.1, had some value.

From these data, the MRSECs compare favorably in citations per publication with the MPIs. The library scientists who compiled the data noted that there was no significant difference between the MRSECs and the MPIs in citations. MPIs are among the premier institutions for materials research in Europe. This cursory survey confirms the previous results of the MITRE report which suggested that citation index comparisons do not sharply distinguish between research results from MRSECs and those from elsewhere; in fact, the MRSECs seem to be just as successful as the German MPI research centers by this metric.

Analysis by Subfields Within the MRSEC Program

The committee also examined the finer structure of the materials research field to determine if MRSECs contributed distinctively in certain subfields—particularly the areas on which the IRGs focused. As part of a survey from each MRSEC, the committee requested a list of the "top 5 scientific questions currently being addressed." Synthesizing those responses and the IRG descriptions, the committee developed the following list of 23 subfields of materials research that may show differential MRSEC impact. The committee does not believe that this list is definitive.

1. Biomaterials
2. Ceramics
3. Composites
4. Ferroelectrics

5. Granular Material
6. Interfaces
7. Liquid Crystals
8. Magnetic Materials
9. Materials for Energy Storage
10. Materials Growth
11. Mesoscopics
12. Mechanical Properties
13. Nanomaterials
14. Nanostructures
15. Organic Semiconductors/Molecular Electronics
16. Oxides
17. Photonics/Optical Materials
18. Polymers (including copolymers)
19. Self-Assembly
20. Spintronics
21. Superconductivity
22. Supramolecular Materials
23. Transport Properties

It is clear that there is a great deal of overlap with the topics covered in the 100 most-cited materials papers and those in the most-cited papers from the MRSECs as shown in Figure 3.4. The committee chose to pursue the research-subfield impact hypothesis in two different ways: an objective analysis relying on publication citations and a subjective analysis using perceived standing from a panel of voting experts.

Using a scientific publication-citation analysis tool (Scopus), the committee identified the top 30 papers since 1995 in each subfield. Nearly 700 papers were selected for analysis. To get an idea of the level of activity in different subfields, the total number of citations for the top 10 papers in each subfield was tabulated (as shown in Figure 3.7). Comparison with Figure 3.4 indicates that there is substantial overlap of the most active areas with the corresponding efforts in the MRSEC program. Figure 3.8 shows a breakdown by country of origin, and Figure 3.9 shows the total.

Further analysis of the data set involved sorting the top-cited papers in terms of the type of institution where the research was performed. As can be seen in Figure 3.10, over the past 10 years it is the universities that have supplied the most highly cited research. This contrasts with the finding from the first look at significant new materials, which showed that they most often originated in industrial laboratories. This may result either from the higher value placed on publications in academe or from the decline of support at many industrial research laboratories.

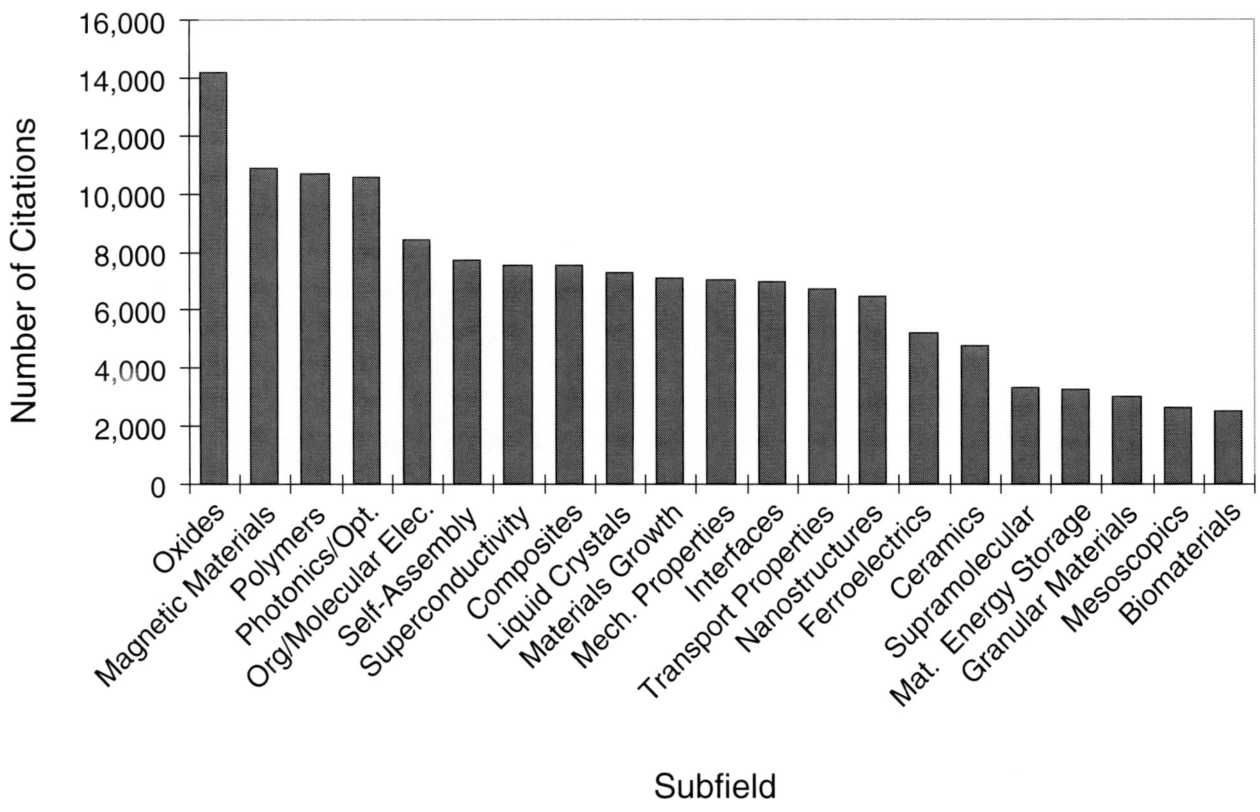

FIGURE 3.7 Total number of citations for the top 10 papers in each subfield.

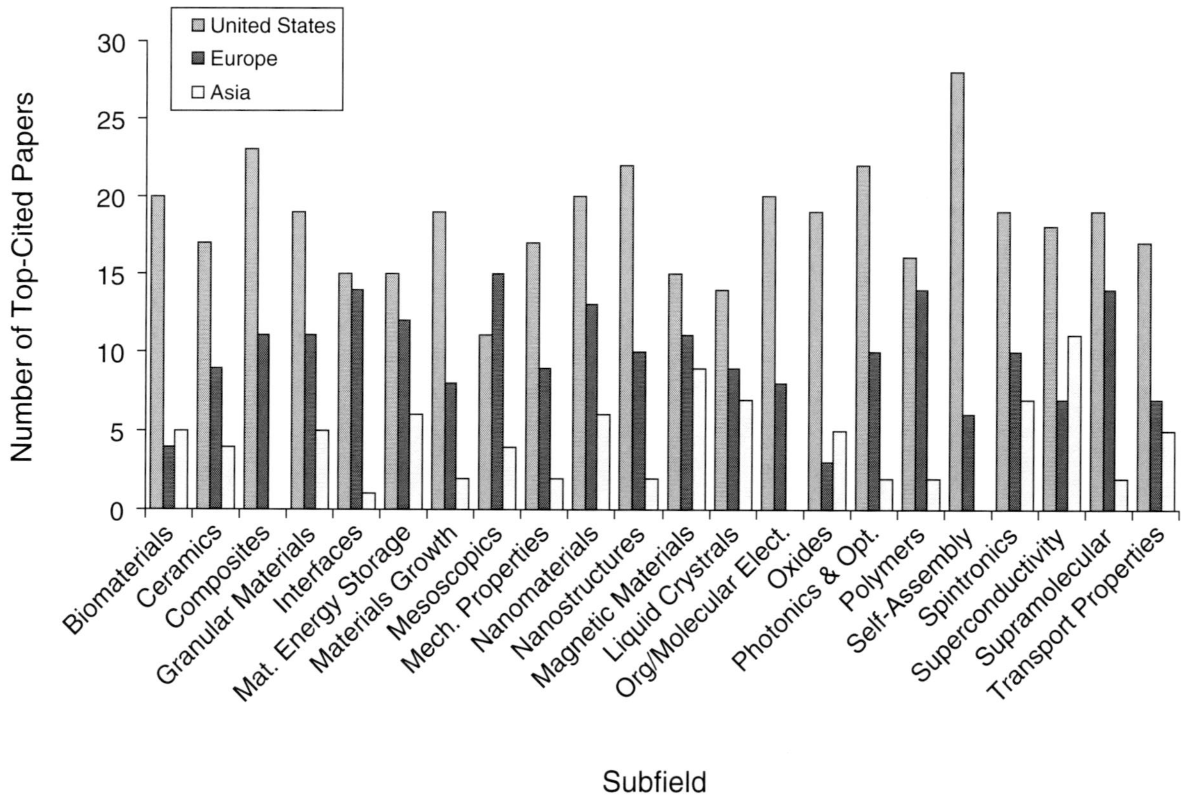

FIGURE 3.8 Top-cited papers in each of 22 subfields of materials research, by world region (United States, Europe, Asia), 1995-2006. The United States remains the largest single source (54 percent) of highly cited papers, followed by Europe (28 percent) and Asia (12 percent). The United States has the most top-cited papers in every subfield except mesoscopics, although Europe has similar numbers in interfaces, energy storage materials, polymers, and supramolecular systems. Top-cited papers from Asia seem to be concentrated most strongly in several areas, particularly the key areas of superconducting and magnetic materials. The integration over subfields is shown in Figure 3.9.

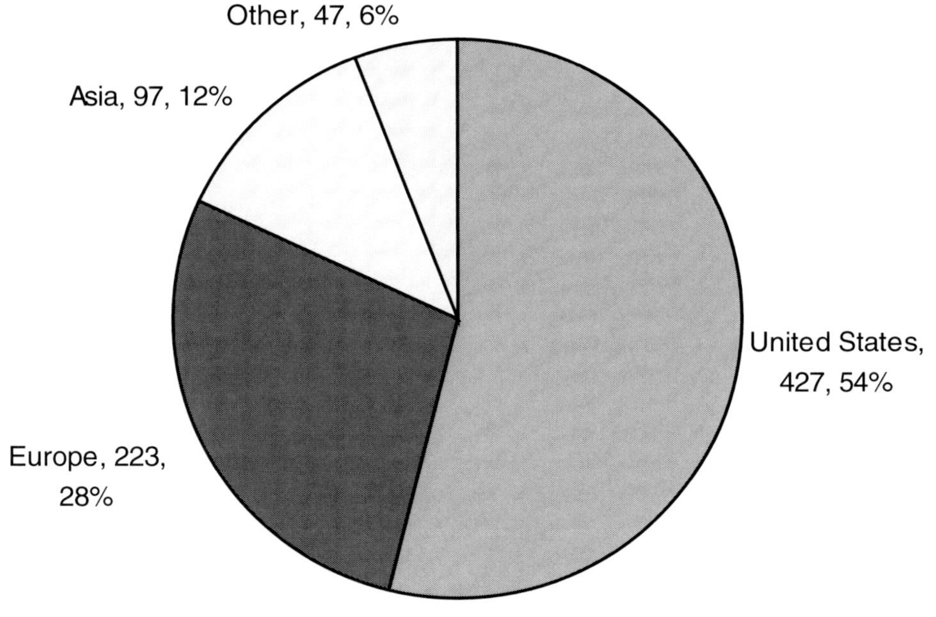

FIGURE 3.9 Geographical region of origin for top-cited papers summed over the 23 subfields of materials research, 1995-2006: number of papers and percentage of total. Eighty-four percent of the papers had all authors from the same region.

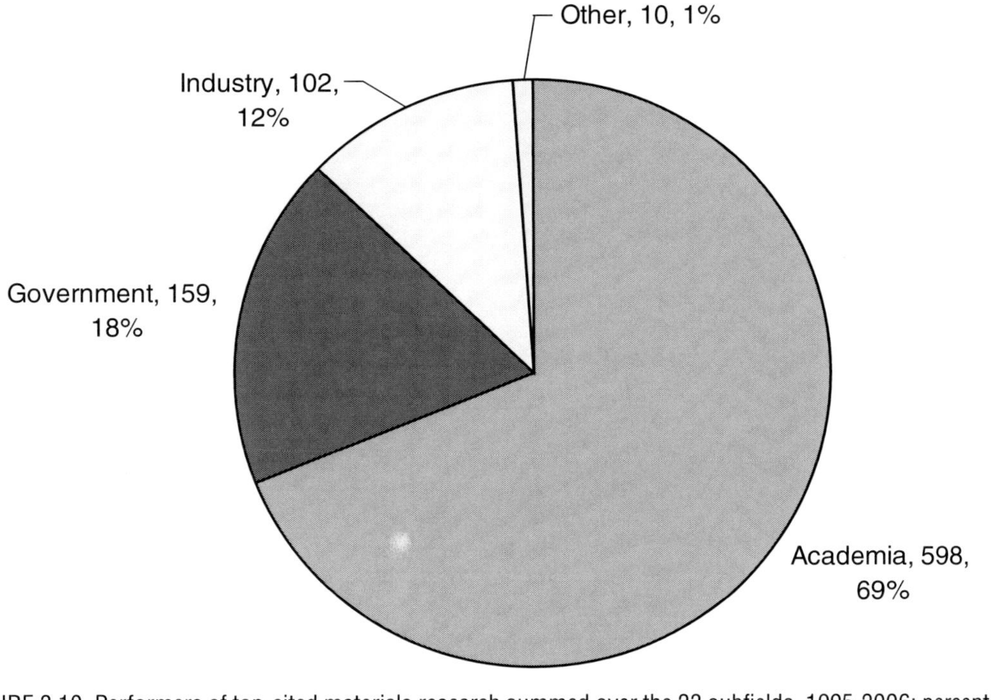

FIGURE 3.10 Performers of top-cited materials research summed over the 23 subfields, 1995-2006: percentage by type of institution.

One of the main objectives was to see whether this study of highly cited work could document that the MRSECs played a substantial role. Figure 3.11 shows the acknowledged funding sources for the most highly cited papers summed over the different subfields. The fraction associated directly with MRSECs is small, 4 percent of the total. Of course the fraction should be considered with respect to the total amount of research money that is provided by the different sources. (A more detailed discussion of the publications vis-à-vis the funding from different sponsors is presented later in this chapter.) However, another factor that must be considered is that most of the papers acknowledge more than one funding source. It is difficult to assign agency ownership to a discovery, a new material, or even just a publication.

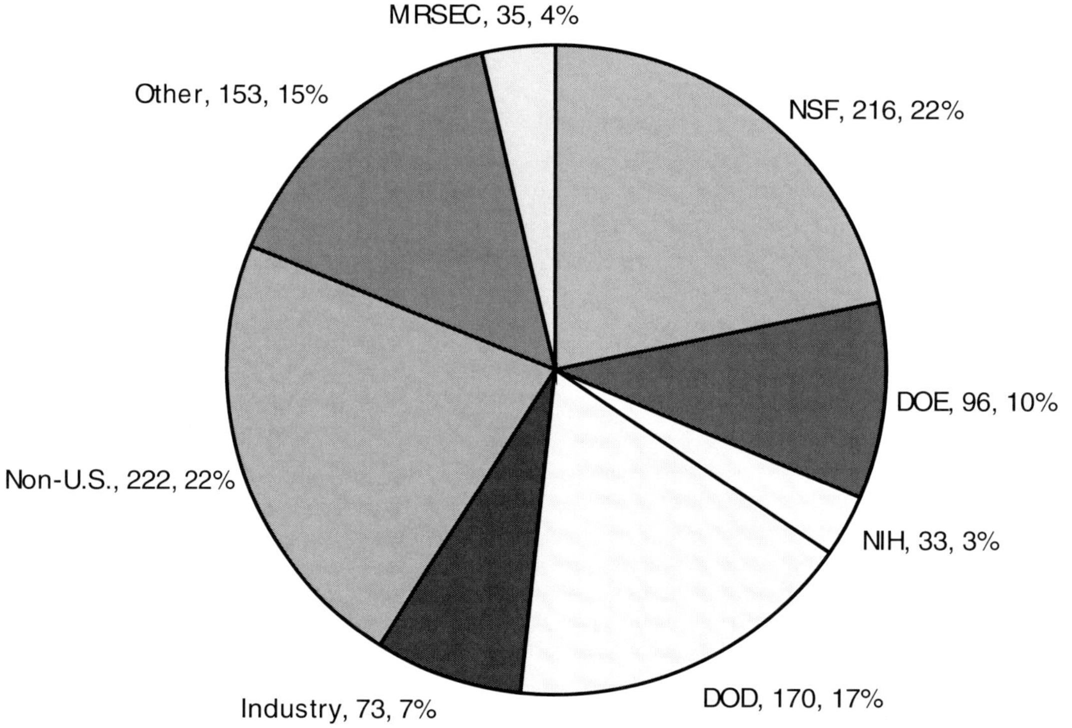

FIGURE 3.11 Sources of funding for top-cited papers from U.S. institutions, summed over the 23 subfields of materials research: number of funding acknowledgments and percentage of total. Papers acknowledging Materials Research Science and Engineering Center (MRSEC) support are counted separately from those supported by other programs at the National Science Foundation (NSF). Seventy-seven percent of the papers acknowledged multiple sources of funding. DOE, Department of Energy; NIH, National Institutes of Health; DOD, Department of Defense.

This analysis also found substantial evidence that MRSEC research is as collaborative as non-MRSEC research. The committee examined the set of MRSEC self-reported top-5 publications. The typical number of senior principal investigators was two per paper, and there were more with single senior authors than with three senior authors. Figure 3.12 compares the proportion of top-cited papers that have single and multiple authors among the different subfields of materials research. While there are some fields in which the top-cited papers are almost evenly divided between single and multiple authorship (i.e., nanomaterials and nanostructures, liquid crystals, organic and molecular systems, photonics and optics, and polymers and supramolecular materials), for the most part papers having multiple senior authors are the norm. Overall, these data indicate that 65 percent of all top-cited papers involve multiple senior authors; however, Figure 3.13 shows that the most likely collaboration is between pairs of senior authors, with vanishing incidence of collaborations with three or more senior authors. There is no evidence that top-cited papers by MRSEC investigators display a different trend, although a primary argument used to rationalize the MRSEC organization is that these larger-scale collaborations are only possible in centers that include a large number of researchers from different departments. Thus, value added by MRSECs in collaborative research is likely on the "input" side (conceptual collaboration in choosing and initiating research directions) rather than the "output" side (research results as measured by published papers).

It is important to note that not all these collaborations are within single institutions. Figure 3.10 showed the distribution of top-cited papers, grouped over all subfields, with respect to the institution of the authors. University-based researchers, including those at MRSEC hosting institutions, are the authors of the highest percentage of top-cited papers worldwide (69 percent), with the remaining 30 percent coming from industry and government laboratories. The committee notes that 44 percent of all top-cited research papers involved collaborations among multiple institutions and also multiple institution types (i.e., collaborations between universities and national laboratories, industry and university, and so on). The committee sought also to quantify the prevalence of international collaboration among these top-cited papers. It was found that collaboration remains largely confined to individual countries, with only 16 percent of the papers involving international collaboration.

In summary, the committee set out to establish the baseline publication-citation characteristics of the general materials research community in order to enable a comparison with the self-reported MRSEC publications. However, in so doing, the committee came to realize that distinguishing MRSEC-enabled research papers was much harder than imagined. As Figure 3.11 shows, identifying a "MRSEC" paper is a subjective assertion about "whose dollars" put the research task over the

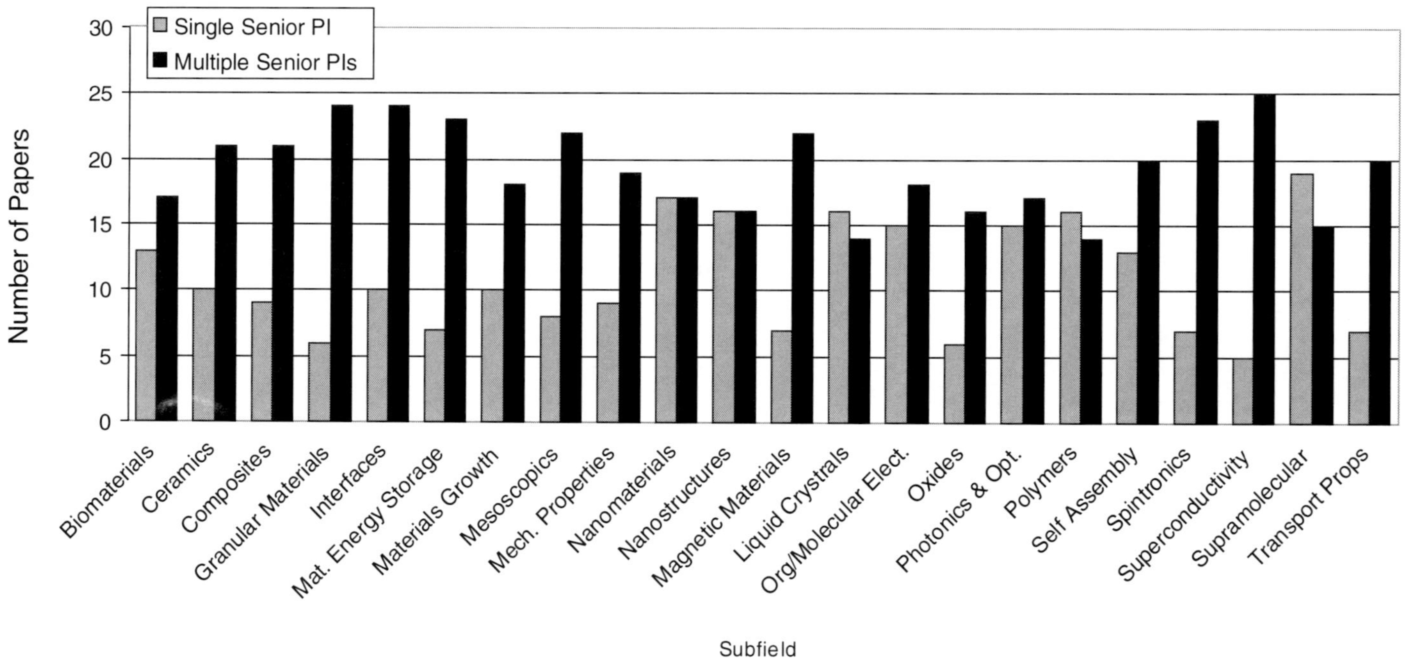

FIGURE 3.12 Comparison of the number of top-cited papers by single or multiple senior principal investigators (PIs) among the different subfields of materials research. Sixty-five percent of the papers had multiple PIs.

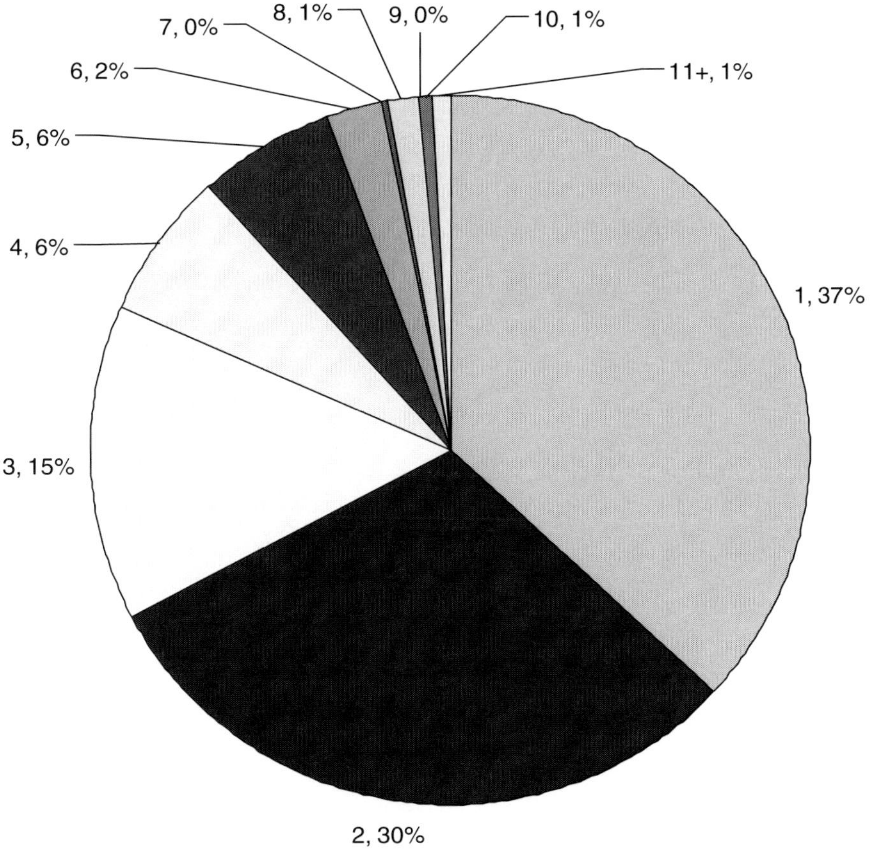

FIGURE 3.13 Proportion of top-cited papers with number of senior authors indicated, 1995-2006, summed over the 23 subfields. The key result is that more than half of the papers were authored by at least two senior principal investigators.

tipping point. The committee did not conduct a fully parallel analysis of the self-reported MRSEC papers as a result.

DEMOGRAPHICS OF RESEARCH PERFORMERS

A primary objective of the original MRL program that has continued into the MRSEC program is to provide a setting that stimulates and nurtures interdisciplinary collaborative materials research. The original MRLs were located at institutions that already had a substantial interdisciplinary materials effort. Figure 3.14 shows the mix of disciplines in the first MRLs. In the original 10 MRLs, about 26 percent

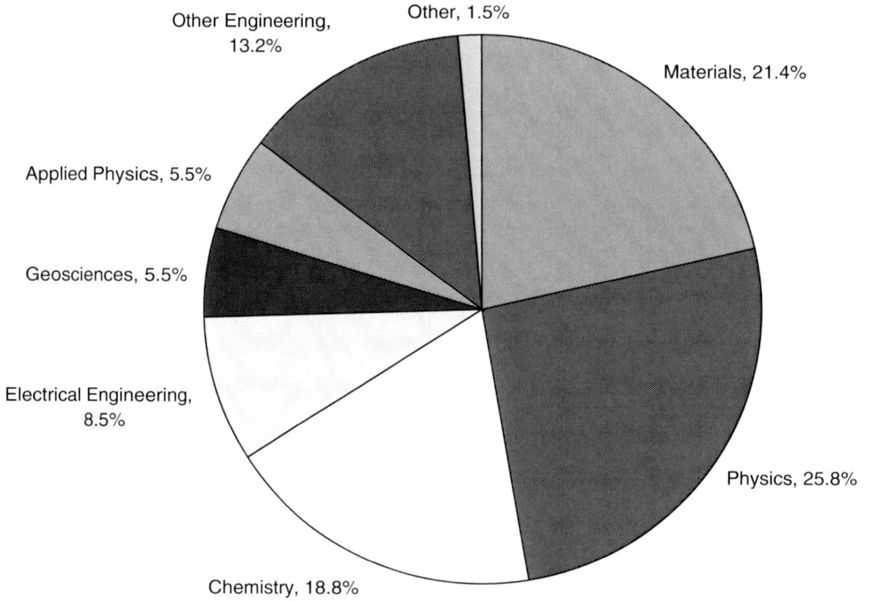

FIGURE 3.14 Reported disciplinary affiliations for participants in the Materials Research Laboratory program in the mid-1970s. SOURCE: MITRE Corporation, *Evaluative Study of the Materials Research Laboratory Program,* 1976.

of the researchers came from physics departments, with a similar number from materials science and engineering departments and chemistry departments. Similarly, about 23 percent of the participants were from other engineering departments (including electrical engineering). It is interesting to note from Figure 3.15 that this departmental mix is almost the same in current MRSECS, with the only meaningful change being a modest increase in the participation from physics departments. A generation after the establishment of the MRLs, there is no question that MRSEC research remains both broad and multidisciplinary, and perhaps one can make the argument that this intrinsic attribute of the MRL and MRSEC programs has led the trend in materials research more generally.

The committee finds, therefore, that on the metric of multidisciplinarity as measured by departmental affiliation on research papers, the MRSEC program performs similar to the overall materials research community. However, the committee did not examine the paper-by-paper distribution of departmental affiliations. And, as before, it could not find a clear way to measure what the degree of multidisciplinarity would have been in the absence of the MRSEC program.

Figure 3.16 shows the distribution of departments of authors of top-cited papers from the committee's list of the most-cited papers by subfield. The distribution

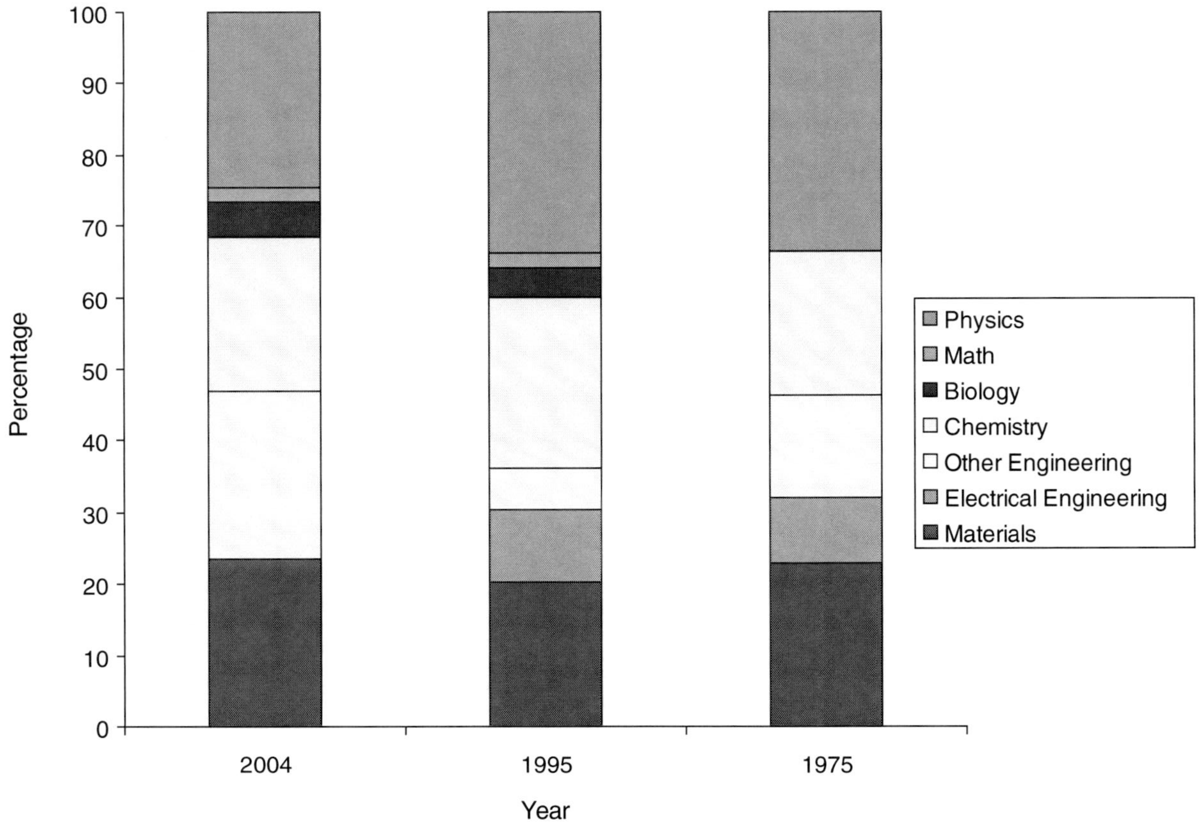

FIGURE 3.15 Individual academic departments by discipline as a percentage of total departments participating in the Materials Research Laboratory and Materials Research Science and Engineering Centers programs. The most recent information, from 2004, is on the left-hand side of the figure. The order of departments from top to bottom on the chart is shown in the legend.

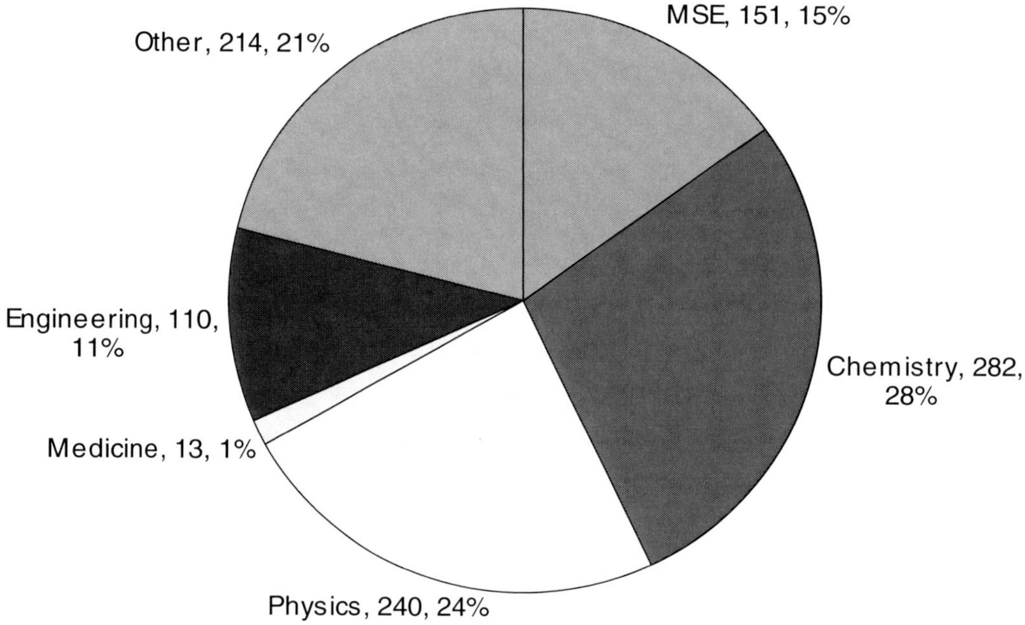

FIGURE 3.16 Departmental affiliations of authors of top-cited materials papers, summed over the 23 subfields, 1995-2006: number of papers cited and percentage of total. NOTE: MSE, materials science and engineering.

is similar in many ways to the departmental mix in MRSECs, with almost identical representation of physics and chemistry in the two distributions and a somewhat weaker participation of materials science and certainly overall engineering in the author distribution. Assuming that the likelihood of a field producing a top-cited paper increases when there are more researchers in the field, these data suggest that the mix of disciplines represented in current MRSECs is similar to the mix found in the worldwide materials research community, which has become deeply interdisciplinary.

A modest longitudinal study was also performed to see how the division of top-cited papers among different types of U.S. and foreign research organizations has evolved over the period of time during which MRSECs and MRLs have been active. Here, the data were not broken out into subfields; instead, the 100 top-cited papers in materials research over each of the decades 1965-1974, 1975-1984, 1985-1994, and 1995-2006 were selected. This follows the initial survey in which the very general criterion "materials research" for the 100 most-cited papers was used. It is unlike the previous citation results, which were broken down into MRSEC subareas.

As shown in Figure 3.17, the percentage of top-cited papers is plotted for U.S.

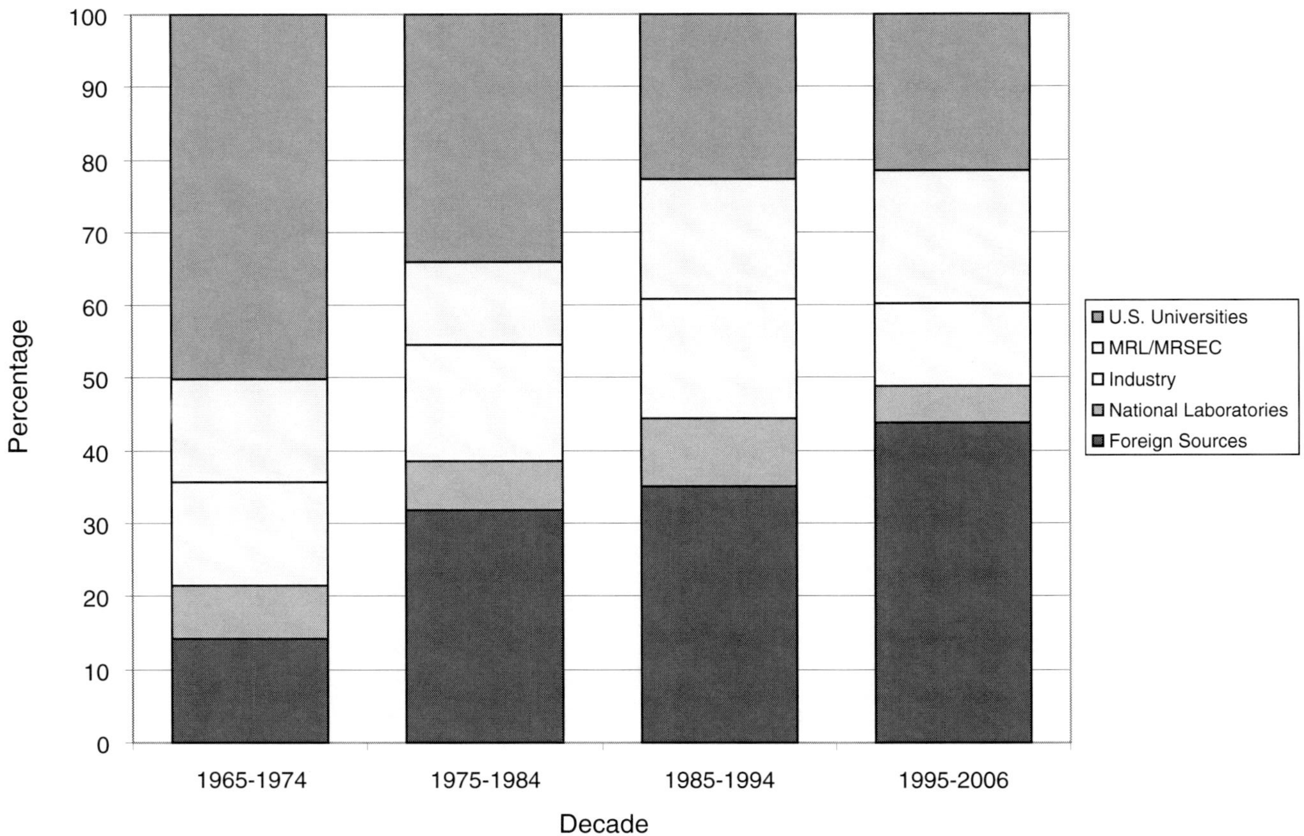

FIGURE 3.17 Sources of the top 100 most-cited materials papers as a percentage of total for four decades, 1965-2006. The order from top to bottom on the chart is shown in the legend. NOTE: MRL, Materials Research Laboratory; MRSEC, Materials Research Science and Engineering Center.

universities, universities having a MRSEC or an MRL, national laboratories, and industrial laboratories, as well as the total foreign citations for each of the decades in Figure 3.17. (Similarly, the number and percentage of top-cited papers in each of the 23 subfields, since 1995, by region of origin, were shown in Figures 3.8 and 3.9.)

It is clear that the United States enjoyed a near-monopoly on top-cited materials papers in 1965-1974 but that this percentage has fallen steadily in subsequent decades to its current level of 54 percent as foreign governments invest in the creation of their own materials science and engineering (MSE) knowledge base.[7] Considering that it may take a decade or more for citations to develop fully for pathbreaking work, this plot is a cause for concern. The percentage of top-cited work has remained roughly constant for U.S. national laboratories, industrial laboratories, and perhaps has even grown slightly for MRSEC-hosting universities over this 40-year period. Yet it is clear that the growth of top-cited papers from foreign institutions has largely been at the expense of top-cited papers from U.S. universities, which in the 1960s produced half of the top-cited work. Ensuring the strength of university-based materials research is crucial, not only because it is the single largest sector of materials researchers in the United States (see Figure 3.18), but also because this is where future generations of materials researchers—both domestic and foreign—will be trained.

THE LEADING GROUPS IN MATERIALS RESEARCH

To assess the perceived excellence of the programs in 23 different subfields of materials research that may show differential MRSEC impact, the committee undertook an informal survey of the opinions of experts. It was proposed initially that the experts be selected by choosing the senior authors of the top 10 most-cited papers from each of the 23 subfields; however, the committee decided that the list should be augmented by authors of highly cited papers who were not selected by the simple algorithm above. For the purposes of this exercise, "expert" is defined as one of the senior authors of one of the 10 most highly cited papers in each of the 23 subfields listed above. The experts were then contacted by e-mail with the following sample note:

Dear Dr. X,

We're working on a National Research Council report on materials programs in the US. As part of the evaluation we thought it would be useful to find out where the best research is being done. To wit, we have identified a set of experts in materials related subfields and would like to solicit their opinions. We would therefore greatly

---

[7]National Research Council, *Globalization of Materials R&D: Time for a National Strategy*, Washington, D.C.: The National Academies Press, 2005, p. 2.

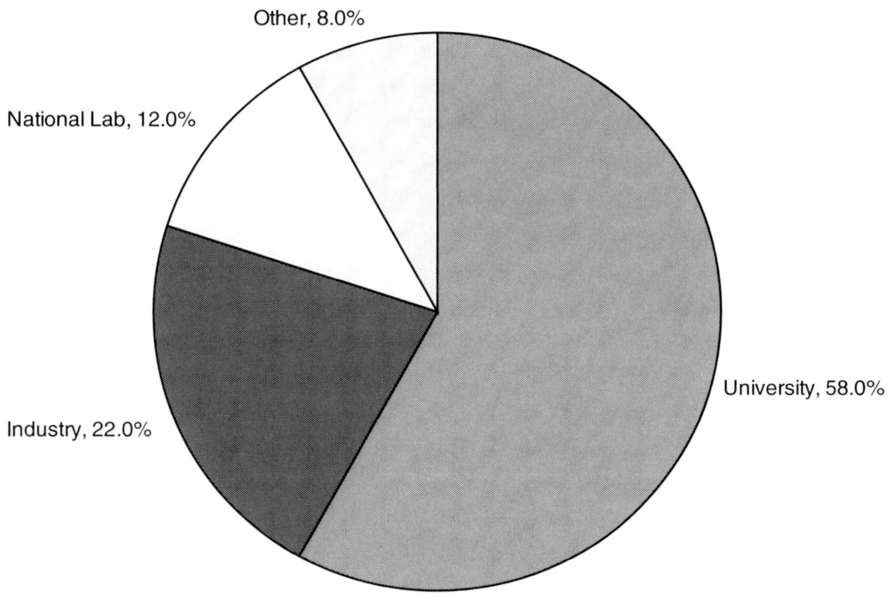

FIGURE 3.18 Affiliation of U.S. materials research community as estimated by demographics analysis of members of the Materials Research Society. The committee notes that the membership of the Materials Research Society is not broadly reflective of the overall composition of the materials research community, but it does have certain parallels to the university-based research community of Materials Research Science and Engineering Centers. (Courtesy of the Materials Research Society.)

appreciate your expert opinion of the top research labs (~10), world-wide, in the area of "granular materials." Thank you for your help.

Sincerely,

NRC MRSEC Impact Assessment Committee

About 200 e-mail inquiries were sent, and 55 experts replied with lists. Several were experts in more than one field and provided several lists. It was not possible to meaningfully rank institutions in each subfield on the basis of these data. By combining the subfields, however, the committee found sufficient evidence from which to draw conclusions as to the reputation of different institutions in the overall area of materials research. The responding experts were widely distributed in foreign laboratories and universities and domestically in institutions with MRSECs and without them, as shown in Figure 3.19.

The votes were then tallied for each subfield. Almost all respondents sent "top 10 lists" with a note that the institutions were not in order of excellence. A vote was counted each time that an institution was mentioned on an expert's list. The

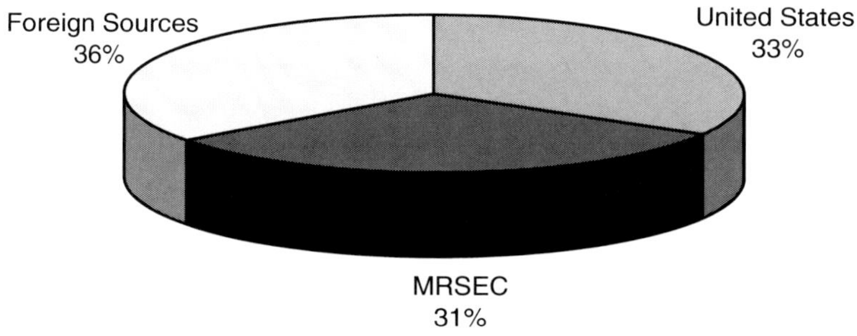

FIGURE 3.19 Sources of "expert votes" in the survey of leading research groups in materials. The "United States" category does not include the votes from institutions with a Materials Research Science and Engineering Center (labeled "MRSEC" in the chart).

main finding of the exercise is shown in Figure 3.20. The institutions were sorted according to their status as domestic university or national laboratory/industrial laboratory, and European, Asian, and other (Canada, India, Israel). The U.S. universities are subdivided further as to whether or not there is a MRSEC on the university campus.

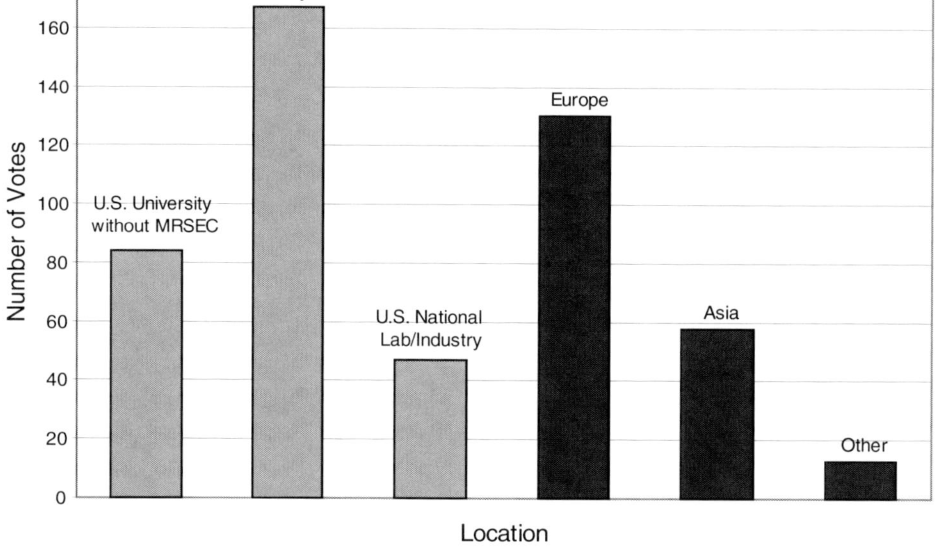

FIGURE 3.20 Location by general categories of the "leading research groups," summed over the 23 subfields of materials research, according to votes by experts. Note that the data in this figure closely match the data in Figures 3.8 and 3.9.

This survey of the most highly regarded research laboratories, across the subfields of materials, documents the leading role played by U.S. universities with MRSECs. As a group, U.S. universities with MRSECs are more identifiable with perceived excellence in materials research than any other grouping in the committee's survey. Beyond the strong correlation of universities with MRSECs and perceived leadership, it is difficult to document whether this correlation is cause or effect. Some of the groups are MRSEC-supported while others are not. None of these world-leading groups is solely supported by a MRSEC. In at least one case, an expert specifically claimed that the institution's MRSEC was not supporting the top-ranked group.

One might be tempted to contrast these results with those in the list of major discoveries described above in this chapter. Note, however, that the major discoveries were dominated by developments from 10 to 40 years ago, while this virtual-voting exercise was sensitive to contemporary impressions of perceived importance. Furthermore, the exercises were sensitive to different characteristics of research excellence: major discoveries versus overall high quality.

The distribution of leading groups was not uniform across the MRSEC program. Of course, there is some correlation between the number of years that an institution has had a MRSEC and its funding level and how well it does on this plot. But again, it is difficult to draw direct conclusions other than that MRSECs are situated at places that do excellent materials research.

RESEARCH IMPACT VERSUS FUNDING: QUALITY PER DOLLAR

Figure 2.2 in Chapter 2 showed the total federal funding for basic materials research between 1982 and 2002. The as-spent funding for materials research was almost constant in the 1980s, although the decade was followed by growth of about 35 percent between 1994 and 2000, in part reflecting the broadening of fields considered to be "materials research."

Federal agencies support materials research at basic, applied, and developmental levels. When all of these expenditures are aggregated, the total exceeds \$2 billion in FY 1995. The last published summaries of these expenditures were made for FY 1994 by the Materials and Technology Committee (MatTec), a subcommittee of the Federal Coordinating Council for Science, Engineering, and Technology reporting to the Office of Science and Technology Policy. It is important to note that virtually all major federal agencies supporting research are represented in this total. Figure 3.21 shows how the \$2.124 billion total in 1994 was apportioned among the different agencies. NSF accounted for about 16% of this total, with about half (56%) of this total coming from DMR. (Note: subtracting facilities, \$288 million for the Department of Energy [DOE] and \$28 million for NSF, one gets a total of \$1.793 million, with NSF at 15% and DOE at 34%.) The MRSEC percentage was

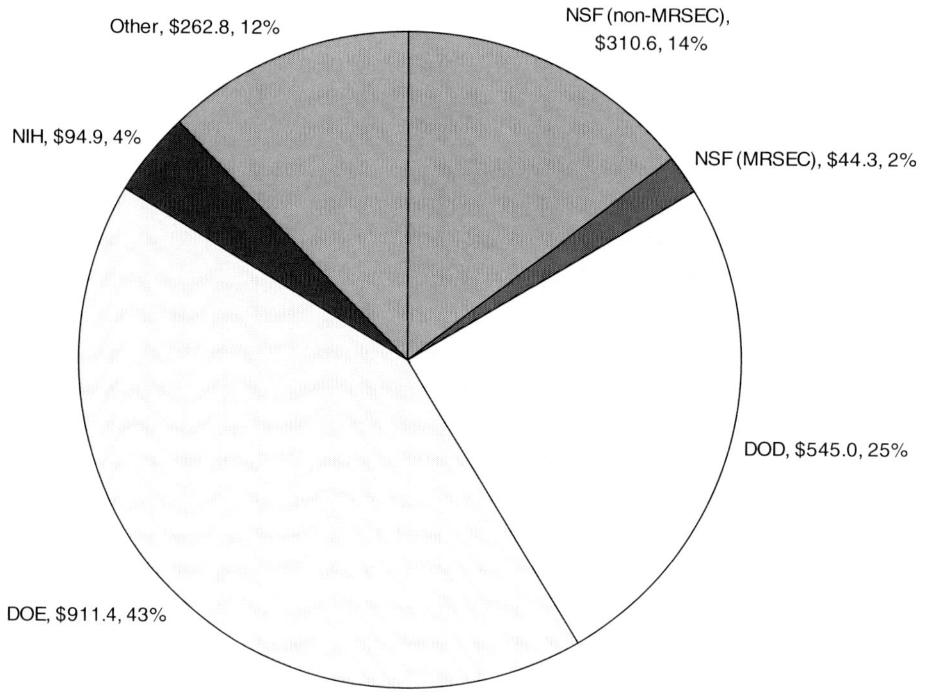

FIGURE 3.21 Expenditures by federal agency for basic, applied, and developmental materials research and engineering in millions of U.S. dollars and as percentage of total expenditure ($2,124 million) in FY 1994. The data do not include classified research or the construction and operating costs associated with facilities. SOURCE: Materials Technology Subcommittee, National Science and Technology Council, *1994 Annual Report*, 1994.

about 25% of the total DMR expenditure, very similar to 2006 proportions. Altogether, MRSEC expenditures represent a very small fraction of the federal materials portfolio, amounting to about 2% of the total.

Of course, given the variety of activities funded by this portfolio and the different programmatic needs of the different agencies, there is no a priori reason to believe that the number of top-cited papers claimed by a given funding agency is proportional to its relative level of materials funding. The acknowledged sources of funding in the top-cited papers are shown in Figure 3.11. It is important to note that fully 77% of all papers acknowledged multiple sources of funding, implying that—increasingly—no one agency can take sole credit for funding any piece of work. There is no simple relationship between the level of federal funding and the percentage of top-cited papers enabled by this funding. For instance, DOE provides 43% of the support for all basic materials research and garnered 10% of the top-cited papers, while NSF provided 16% and was acknowledged in 16% of the top-cited papers. These data are compiled in Figure 3.22, which compares the percentage of top-cited papers from 1996 to 2006 that acknowledge each agency with the percentage of the total federal budget for materials research spent by that agency.

While the overall monetary investment is very different, for the National Institutes of Health (NIH) and the Department of Defense (DOD) there is good agreement between the percentage of the top-cited papers and the percentage of the federal budget used to enable the research in these papers. The NSF represents relatively good value for investment, yielding top-cited research papers at almost twice the rate per dollar invested. The MRSEC program (2% of total materials investment, 5% of top-cited papers) is similar in "efficiency" to NSF overall (14.6% of total materials investment, 30% of top-cited papers). DOE has very high research expenditures but a relatively lower participation in top-cited papers. This is probably because the embedded cost of constructing and operating the DOE user

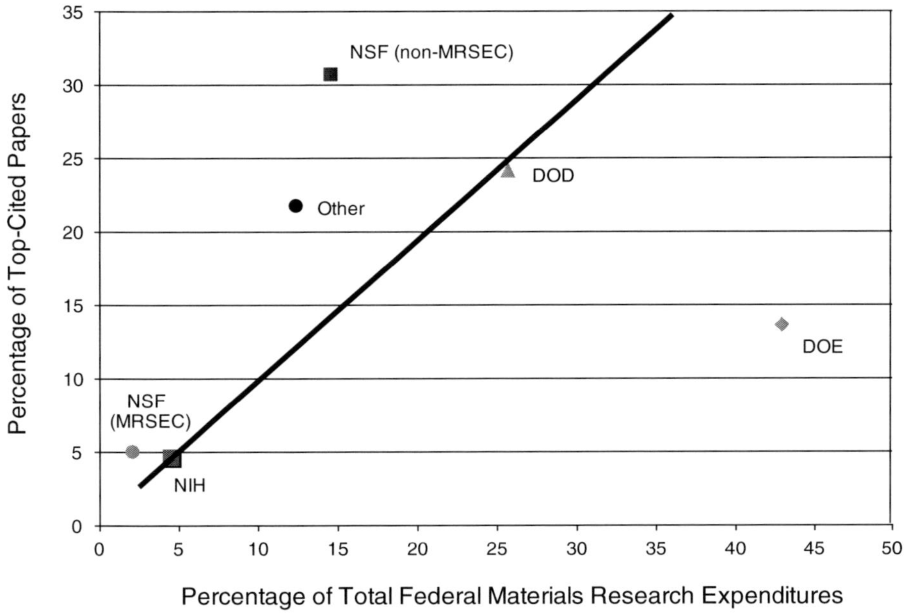

FIGURE 3.22 Comparison of number of top-cited materials papers to materials research expenditures for major federal agencies, 1996-2006. The solid line has a unit slope. The committee was not able to easily separate the university and national laboratory components of the DOE materials research budget; therefore, the DOE data point is not consistent with the university-research program budgets used for the other agencies. NOTE: NSF, National Science Foundation; MRSEC, Materials Research Science and Engineering Center; DOD, Department of Defense; DOE, Department of Energy; NIH, National Institutes of Health.

facilities is not properly accounted for. The committee also notes that in selecting the data for the plot in Figure 3.22, the papers originating only from DOE laboratories were excluded in order to allow a comparison of DOE-supported university research with NSF-supported university research. Including the DOE national laboratories more than doubles the number of such papers. It can be concluded from Figure 3.22 that NSF research, including that carried out in MRSECs, is more likely to result in a top-cited publication than research funded by any other major agency. Within these statistics, there is little evidence that the MRSECs are more or less productive in this respect than any other NSF materials program.

SHARED EXPERIMENTAL FACILITIES

An often-cited key element of the MRSEC program is its explicit provision of shared experimental facilities (SEFs) at each center. The MRSEC program provides only limited explicit support for underwriting the capital costs of acquiring and maintaining a comprehensive instrument suite; rather, institutions must find other mechanisms for purchasing equipment (including the use of other NSF programs). MRSECs SEF funds, originating from budgets for IRGs, seeds, and facilities, are usually expended to cover operating costs of equipment and facilities such as maintenance, supplies, or portions of a salary for technical support staff (see Figures 3.23 and 3.24 for the distribution of MRSEC SEF budgets compared to each center's budget). In 2004, DMR estimated that 12 percent of the MRSEC budgets was spent on capital equipment.

The research and training of students and postdoctoral associates in the MRSECs is completely dependent on the availability of SEFs with forefront capabilities. The MRSEC SEFs support a very broad range of materials research, which is essential to a broad community (including many supported by single-investigator grants) but is not altruistic, since the MRSECs could not carry out their own research without the user fees generated by these users.

As identified in the National Research Council (NRC) report *Midsize Facilities: The Infrastructure for Materials Research*,[8] each materials research facility secures its capital and operating sources of support in a unique and highly individualized fashion. NSF MRSEC SEF support is often only one component of a complex array of funding mechanisms. Many MRSECs operate their SEF facilities with some user fees in order to recover some of the operating costs. In the larger MRSECs, the SEF user community is larger than the number of MRSEC students by at least a factor of 10. This large user base is necessary to pay SEF staff salaries that could not be sustained on the MRSEC budget alone. Another common feature was that

[8]National Research Council, *Midsize Facilities: The Infrastructure for Materials Research*, Washington, D.C.: The National Academies Press, 2006, p. 62.

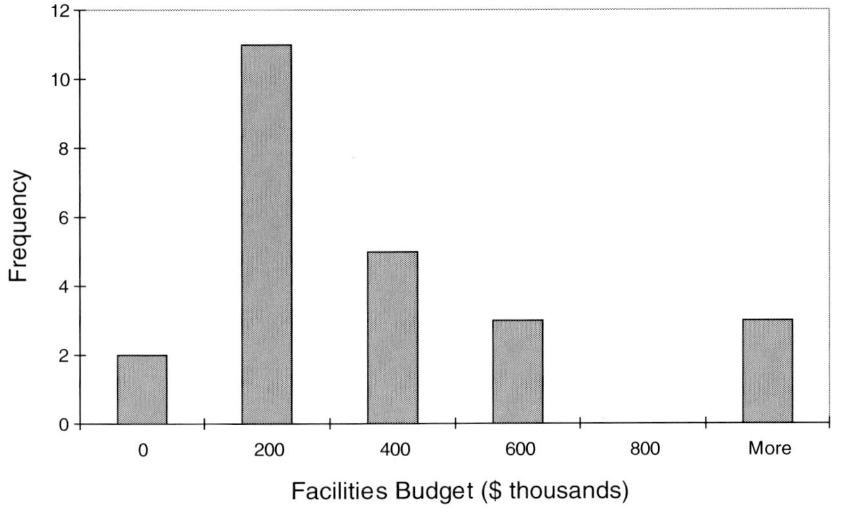

FIGURE 3.23 Distribution of Shared Experimental Facilities budgets for most of the Materials Research Science and Engineering Centers in 2005-2006.

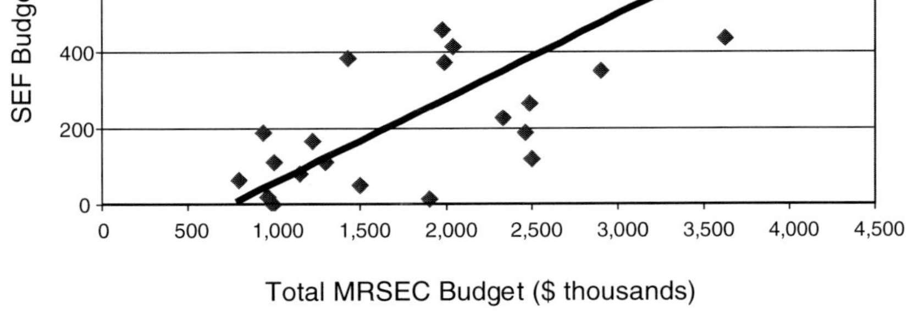

FIGURE 3.24 Correlation plot of total Materials Research Science and Engineering Center annual budget versus Shared Experimental Facilities budget. As expected, the correlation is positive but not linear.

faculty and students who participated in the MRSEC would receive slightly discounted rates for using the instruments as compared with other users on campus. It is also important to note that in most cases, the instrumentation supported under the MRSEC SEF program element was part of a larger suite maintained by the institution.

In terms of impact, the committee believes that the shared facilities supported by MRSECs do have significant impact in the larger community, but the committee was not convinced that the MRSEC SEF support was dramatically more effective or leveraged more than any other instrumentation program. For instance, the committee learned from the *Midsize Facilities* report referred to above that operating costs for shared facilities (including the MRSEC program) are recovered about equally from federal grants, user fees, state awards, and institutional commitments. In examining the MRSEC annual reports, the committee observed a similar mix of reported sources of operating costs for the SEFs. However, the committee was unable to collect reliable data about sources of funds for the acquisition of capital equipment both inside and outside the MRSEC program. It is the committee's view that MRSEC centers likely attract elevated levels of cost sharing from institutional leaders because they attract attention and provide explicit federal leveraging. The committee notes again that the specific impacts are probably diluted when viewing average trends. For instance, MRSEC participants at the University of Southern Mississippi credited the MRSEC with helping empower them to compete successfully for additional instrumentation awards from NSF and other agencies. The committee did not measure and compare the degrees of utilization of facilities inside and outside the MRSEC program and therefore cannot comment on the relative accessibility of the instrumentation to the broader community. The general perception seems to be, however, that MRSECs do allow wide-ranging access to their facilities.

The committee found that MRSECs invest in facilities at a rate comparable to DMR overall and that MRSECs provide about 20 percent of the DMR instrument portfolio. The committee also observed that—averaged over the past 10 years—institutions with MRSECs attracted "instrumentation for materials research" awards at roughly the same rate as institutions without MRSECs. The committee could not easily measure, however, whether institutions with MRSECs attracted a higher volume of instrumentation awards from sources outside NSF.

The committee collected information on the levels of MRSEC support that were directed toward shared experimental facilities from the annual reports of the MRSECs. The following observations were made.

- In 2004, the average MRSEC budget spent on facilities was \$276,000 (median \$187,000) per year with a total reported investment of \$6.6 million;

$6.6 million is about 13 percent of the annual $50 million MRSEC program budget.

- In 2004, the portion of the DMR budget spent on equipment and instrumentation was $30 million (beyond that of the MRSEC program), or about 12 percent of the division's full budget. In addition, DMR distributed about $5.7 million in equipment and instrumentation funds through the Instrumentation for Materials Research (IMR) program. Thus, in 2004, DMR invested about 18 percent of its annual budget (excluding the MRSECs) in equipment and instrumentation.

The committee observes then, that MRSECs invest in facilities and equipment at a rate similar to the overall DMR portfolio of investments. This analysis is extremely informal. It should also be noted that the committee did not compare the type of instruments bought through MRI and IMR awards and those secured through and for MRSECs. Another estimate suggested that MRSECs house about 20 percent of the overall federal investment at universities in million-dollar-class instrumentation for materials research. Also, it should be noted that SEFs in the materials area are not unique to MRSECs, but at institutions with larger MRSECs the SEFs are usually managed and operated by the MRSEC. If MRSECs did not do this, DMR would need to create some other strong facilities program to support materials research.

The MRSEC facilities budget also supports (at least in part) technical staff members, who train students and maintain the equipment. About $240,000 per year is spent on capital equipment. Estimating that half of the equipment purchased through the NSF instrumentation programs (DMR's Instrumentation for Materials Research program or NSF's agency-wide Major Research Instrumentation program) within DMR ends up in a MRSEC facility, another $5 million—or an average of about $200,000 per center—is added to this amount. Assuming a 10-year life for forefront materials characterization equipment, a center might thus afford a total inventory of equipment of about $4.4 million.

The committee also examined the potential correlation between MRSECs and instrumentation funding.

- In the timeframe 1995-2006, the DMR IMR program awarded about $75 million of grants for the acquisition and development of instrumentation for materials research.
- Of these awards, 30 percent (by dollar value) were made to institutions with MRSECs (that were active at the time).
- MRSEC institutions received $402 million during this time frame, about 39 percent of the total $1.04 billion or so awarded by DMR to all institutions (excluding the $544 million for MRSEC funding).

- Thus, the committee observes that institutions with MRSECs attract IMR awards roughly in proportion with their level of materials research activity (measured by DMR funding levels).

MRSECs are, however, taking a lead in working together on facilities rather than competing with one another. Led by the MRSEC at the University of California at Santa Barbara, the University of Southern Mississippi MRSEC, in collaboration with those at the University of Minnesota and the University of Massachusetts at Amherst, proposed to NSF to create a national facilities network. The award has recently been funded and will be used to encourage off-campus users to take advantage of the facilities; it also helps send students and faculty among the four sets of facilities at "internal" user rates.

The variations in actual capital spending equipment from one MRSEC to another are considerable because the availability of resources hinges on other features of the institution such as the development office, relationships with corporate sponsors, and so on. The recent National Research Council report on shared experimental facilities found that most SEFs that serve the large majority of the materials community have a $1 million to $50 million replacement capital cost, with an average of about $10 million.[9] In fact, the U.S. investment in such facilities is currently well below the replacement level,[10] estimated to be on the order of several billion dollars per year. At present, other sources of support for SEF equipment (typically, the universities themselves, or in some cases foundations) are not large enough to make up the difference in needed support. Thus, the average age of equipment in SEFs continues to increase, with many individual items being more than 20 to 25 years old.

FINDINGS AND RECOMMENDATIONS

Conclusion: Consistent with previous analyses, the committee found no simple, quantitative, objective measure to clearly differentiate the MRSEC research product from that of other mechanisms supporting materials science and engineering research.

The committee found the task of evaluating the impact of MRSEC research quite daunting, primarily because research papers published in peer-reviewed

[9]National Research Council, *Midsize Facilities: The Infrastructure for Materials Research*, Washington, D.C.: The National Academies Press, 2006, p. 2.

[10]National Research Council, *Midsize Facilities: The Infrastructure for Materials Research*, Washington, D.C.: The National Academies Press, 2006, p. 113.

journals rarely attribute the results to a single support mechanism. Moreover, any research, even by an individual researcher associated with a MRSEC, is a combination of activities supported "inside" and "outside" the MRSEC. Thus, even if MRSECs have played a unique role in the research enterprise, such as in enabling the formulation of research projects that could not otherwise have been envisioned, there is no easy way to provide substantiation. It could be that the research enterprise has evolved over the past decade, leading to greater convergence and overlap between MRSECs and other research practices. Thus, it is not currently possible to distinguish the unique contributions of MRSECs.

General Finding: Sponsors of research are increasingly unable to claim "sole ownership" of research results; MRSECs are no exception.

Most research publications now acknowledge multiple sponsors. It is not possible to demonstrate that the MRSEC support yields leadership in discoveries, publications, or citations. In part this is because funding per MRSEC has decreased significantly in the past decade, so that each group requires multiple sponsors.

General Finding: Most highly cited publications contain one or two senior authors, indicating that the size of research collaboration is usually small.

Although the materials field is highly collaborative and the general belief is that the community benefits from interactions between local groups of many individual investigators in the same field, discoveries and publication records indicate that over 50 percent of the published papers are from individuals and groups of two.

Although the committee was unable to identify MRSEC-enabled research in "blind taste tests," it successfully assessed the overall research quality in comparison with the research enabled by other mechanisms and elsewhere around the world. For instance, do published research results that acknowledge MRSEC resources achieve citation indices and other measures of impact comparable with research enabled by individual-investigator awards?

Conclusion: Overall, the MRSEC program produces excellent, frontier science of the same high standard as that supported by NSF through other mechanisms. In terms of quality, MRSEC research is at least on a par with that of other multiple-principal-investigator programs and individual grants in the United States and internationally, and is an important element of the overall mix for support of materials research, including support for big centers and single-investigator grants.

- The outstanding discoveries, leading research groups, and most significant publications worldwide are associated with universities at which there are MRSECs.

- MRSECs are involved in the most active areas of materials research as established by their publication records compared with those of the entire field.
- The MRSEC program has the same level of research collaboration as found in comparable national and international groups.

The committee studied a set of major breakthroughs in materials research over the past four decades. U.S. universities, and in particular MRSECs and their predecessors the MRLs, played a limited but pivotal role in a handful of these discoveries. The committee conducted several comprehensive analyses comparing citations of MRSEC-reported research publications and those of the broader research community. The distribution of MRSEC-reported "top-cited papers" across subfields of materials research was very similar to that of the top 100 most-cited papers. Affiliations of the top 100 research papers also showed a 10 percent contribution from institutions with MRSECs or MRLs. The committee also found that the top MRSEC papers were cited much more frequently than the average materials research paper but that the best-of-the-best materials research papers had significantly more citations. However, these papers generally predate the emergence of the MRSEC program. The committee also found that the MRSEC program has the same level of collaboration as that found in comparable national and international groups. To some extent this may be the ultimate success of the MRSEC program in having fostered this type of research at an early stage. Finally, the breakdown of departmental affiliations of MRSEC authors and those of the top-cited materials research papers were quite similar.

In two related exercises, the committee examined the global stature of MRSEC-related research groups. In comparison to the Max Planck Research Institutes of Germany, the MRSECs' publication-citation rates were quite comparable. In a "virtual voting" exercise, the committee contacted researchers around the world in several different subfields and solicited their opinions about world-leading research teams. Research teams at institutions with MRSECs dominated the results.

Although many of these measures are of correlation and not causation, the committee came to believe that the research program enabled by MRSEC awards has been, in general, at least as effective as that enabled by other mechanisms.

Conclusion: The MRSEC program offers one of the principal opportunities in materials research to support shared experimental facilities (SEFs) that include not only equipment but also the personnel to provide training for students and to perform maintenance. Growing constraints on the per capita MRSEC budget have greatly diminished this ability, which is a concern for the infrastructure of materials research in general.

It should be noted that SEFs in the materials area are not unique to MRSECs, but at institutions with larger MRSECs the SEFs often are managed and operated by the MRSEC. If MRSECs did not do this, DMR would need to create some other strong facilities program to support materials research. A large user base is necessary to pay SEF staff salaries that cannot be supported solely by the MRSEC budget. The MRSEC SEFs support a very broad range of materials research (and sometimes other kinds), which is essential to a broad community (including many supported by single-investigator grants), but it is not just altruistic—the MRSECs could not carry out their own research without the user fees generated by these users. Shared facilities are an important resource for the overall community. For instance, individual investigators are unlikely to be able to afford to acquire and maintain a cutting-edge transition electron microscope, whereas a MRSEC SEF would be ideally suited to do so. Such an instrument sited at a MRSEC would be highly leveraged (because of institutional commitments to existing infrastructure and an established user community that would supply fees-for-use) and would greatly expand the opportunities available to the local research community. The committee encourages recent efforts by the centers and NSF to use modest supplemental grants to encourage and promote broader access to these facilities. These instruments are a core part of the value of the MRSEC program and can have enhanced national impact through improved communication and coordination.

As described in the beginning of this chapter, the committee concludes that the merit of the research enabled by the MRSEC program is comparable with the best of the materials research supported by other mechanisms. The committee notes, however, that it focused on measuring the impact of research results and that the ancillary benefits of MRSECs are not reflected by these metrics.

4

Assessment of the Impact of MRSEC Education and Outreach

Education and outreach (EO) covers a broad range of activities that serve kindergarten through grade 12 (K-12) students and teachers; undergraduate, graduate, and postdoctoral researchers; policy makers; and the public. Consistent with the breadth of activities, EO projects serve many different purposes: educating future scientists and engineers; broadening the participation of underrepresented groups in science, technology, engineering, and mathematics (STEM) disciplines; increasing science literacy in the public; informing the public and policy makers about scientific and technical issues; improving K-12 science education; and developing a scientific and technical workforce.

INTRODUCTION

Although all National Science Foundation (NSF) proposals are required to address the "Broader Impacts" of the proposed research, an EO component is specifically required by the Materials Research Science and Engineering Centers program (MRSEC program) announcement (see Box 4.1). In contrast to the efforts of most individual-investigator and small-group grants, many (but not all) Materials Research Science and Engineering Centers (MRSECs) choose to have at least one part-time person (the EO coordinator) dedicated to managing the EO projects.

NSF does not require MRSECs to conduct specific EO activities, with the exception of participation in the Research Experiences for Undergraduates (REU) program and a requirement for plans to increase the number of people from *underrepresented groups* (defined by NSF as women, Hispanics, African-Americans,

BOX 4.1
The MRSEC Request for Proposals on Education and Outreach

The scope of activities of each Materials Research Science and Engineering Center (MRSEC) depends on the capabilities of the proposing organization. Examination of the 2004 program solicitation from NSF reveals that among the list of activities that most MRSECs incorporate "to an extent consistent with the size and vision of the Center" are the following:[1]

> Programs to stimulate interdisciplinary education and the development of human resources (including support for underrepresented groups) through cooperation and collaboration with other organizations and sectors, as well as within the host organization. Cooperative programs with organizations serving predominantly underrepresented groups in science and engineering are strongly encouraged.

The program solicitation description of what should be included about education and outreach in the proposal is as follows:

> Education, Human Resources Development. Describe the education and human resource goals, provide a rationale for those goals, and indicate desired outcomes for the 6-year period. Briefly describe how the education goals integrate strategically with the research and organizational/partnership opportunities of the Center. Outline plans for increasing the participation of women and underrepresented minorities in Center research and education activities. Outline plans for seminar series, colloquial workshops, conferences, summer school and related activities, as appropriate. Describe any additional education programs not included in other sections of the proposal. Limit: 3 pages.

The program solicitation also specifies that "innovative interdisciplinary educational ventures" are appropriate topics for seed funding.

---

[1] National Science Foundation (NSF), *Program Solicitation for Materials Research Science and Engineering Centers Program,* NSF 04-580, NSF, Arlington, Va., 2004.

and Native Americans/Pacific Islanders) involved in STEM fields. Each MRSEC is encouraged to pursue activities consistent with the research and organizational/partnership opportunities of the center as well as the size and local context of the center.

The committee collected data from a range of sources. Written sources included MRSEC annual reports, program descriptions, MRSEC Web sites, grant proposals, journal papers, and program evaluations. Additional sources included telephone interviews, Research Experiences for Teachers (RET) conference reports, MRSEC EO workshop proceedings, and materials from the National Research Center Educator Network (NRCEN) Web site. A survey specific to EO issues was sent to EO coordinators and MRSEC directors in April 2006 asking for information in order to address issues raised from the preliminary analysis. Information from site visits was combined with data obtained during discussions with many of the MRSEC EO coordinators at a MRSEC directors' meeting in Chicago in April 2006 (see Appendix F for more information).

OVERVIEW OF MRSEC EDUCATION AND OUTREACH ACTIVITIES

The flexibility of the NSF EO guidelines has produced a broad range of MRSEC EO activities. As a group, MRSECs reach many difference audiences, including current and future researchers, K-12 and college teachers, students from K-12 through graduate school and (to a lesser extent) journalists, policy makers, and the public. Most MRSECs have a dedicated EO coordinator. EO coordinators may have backgrounds in K-12 education, a STEM discipline, and/or education research. Many EO coordinators divide their time between MRSEC and other programs with similar missions. Some EO coordinators' salaries come entirely from the MRSEC grant; however, it is not uncommon for part of their salary to be paid by the university where the MRSEC is located and/or by other grants. EO coordinators may be involved in setting goals and priorities for the MRSEC, developing curricular and other materials, establishing and maintaining partnerships, facilitating researcher involvement, obtaining additional funding for EO activities, coordinating with other (internal and/or external) EO programs, and assessing, evaluating, and disseminating research or education-assessment results.

Although the EO coordinator is responsible for organizing EO activities and building infrastructure, researchers play an active role in many EO programs. Some MRSECs require a specific number of hours per year from each MRSEC researcher (which includes undergraduate and graduate students, postdoctoral researchers, and faculty), leading to a wide variety of reported researcher involvement. MRSECs also may provide funding for activities initiated by researchers or EO participants (such as teachers) through mini-grant programs.

MRSEC EO activities can be separated into three general modes of operation:

- *MRSEC-funded activities*, in which the MRSEC takes the primary leadership role and provides the majority of the funding from the MRSEC grant;
- *MRSEC-leveraged activities*, in which the MRSEC has obtained additional funding (beyond the MRSEC grant) for EO projects and provides the primary leadership; and
- *MRSEC-associated activities*, in which MRSEC researchers participate in programs run by other entities. The MRSEC may provide a small portion or none of the funding for the program, but it may contribute significant volunteer time.

Goals of MRSEC Education and Outreach

MRSEC EO goals generally originate during proposal development. The goals reported by EO coordinators fall into four main categories:

1. Preparing the future scientific and technical workforce, including researchers at all levels from high school to postdoctoral researchers;
2. Improving the scientific content knowledge of nonscientists through activities for the public, policy makers, and/or K-12 schools;
3. Improving public attitudes toward science, again targeting both the public and K-12 students; and
4. Broadening participation by increasing the number of women and other underrepresented groups involved in MRSEC activities.

Most of the non-research-oriented programs (i.e., those for K-12 and the public) are driven by local factors, including existing programs and specific MRSEC personnel interests. Although most EO programs have a materials science theme, programs for K-12 students and teachers often focus more broadly on general science and engineering or on the scientific process. The subsections below describe some of the current MRSEC activities designed to address these goals for different audiences.

Goal: Preparing the Future Scientific and Technical Workforce

One of the most important functions of a MRSEC is preparing the future scientific and technical workforce. The majority of MRSEC research, as in most academic environments, is carried out by graduate students and postdoctoral researchers (see Table 4.1, and Figure 2.7 in Chapter 2). Although all research grants train graduate students and postdoctoral researchers, MRSECs have a unique opportunity to help students develop skills that they may not learn working for an individual investigator. Most MRSEC students work in collaborative, interdisciplinary groups and learn to use equipment and techniques in laboratories, and sometimes disciplines, beyond their own. Most MRSEC student and postdoctoral researchers receive mentoring from multiple professors but do not typically participate in other MRSEC EO activities. The preparation of the future scientific and technical leadership in materials research often is not reported formally as an EO component; however, it is a very important function of MRSECs. Specific activities falling under the general goal of preparing the future scientific and technical workforce include the REU and RET programs, described below.

Research Experiences for Undergraduates is an NSF-wide program that provides undergraduates with a paid summer research experience lasting from 8 to 10

TABLE 4.1 Percentages of Women and Underrepresented Minorities (URMs) Working in MRSECs in 2005

MRSEC Demographics		Total	Percentage
REU students	Total	279	
	Women	148	53.1
	URM	95	34.1
Undergraduates (App. B)	Total	218	
	Women	86	39.5
	URM	44	20.2
RET teachers	Total	69	
	Women	33	47.8
	URM	16	23.2
Other pre-college	Total	1,545	
	Women	890	57.6
	URM	251	16.3
K-12 students	Total	8,651	
	Women	3,939	45.5
	URM	2,239	25.9
Undergraduate faculty	Total	31	
	Women	11	35.5
	URM	3	9.7
Graduate students	Total	554	
	Women	149	27.0
	URM	30	5.4
Postdoctoral researchers	Total	164	
	Women	33	20.1
	URM	5	3.1
Faculty	Total	419	
	Women	56	13.4
	URM	14	3.3
Technical support staff	Total	81	
	Women	15	18.5
	URM	3.5	4.3
Nontechnical support staff	Total	41	
	Women	37	90.2
	URM	1	2.4

NOTE: For comparison, Table 4.2 shows the approximate percentages of women and underrepresented minorities in the fields most represented in MRSECs. Note that the data in Table 4.2 are from 2002 for the bachelor's and master's degree data, and 2003 for the Ph.D. statistics. There is some uncertainty in the values in Table 4.1 because MRSECs are not required to report how much of the support came from the MRSEC. For example, the number of graduate students claimed in some reports is much greater than the number that the budget shows could be supported. REU, Research Experiences for Undergraduates; RET, Research Experiences for Teachers.
SOURCE: Data in this table are as reported in annual reports of the MRSECs, Appendixes B and C. The annual reports are uniformly prepared in accordance with guidelines from the National Science Foundation.

weeks. NSF funds REU supplements, which are granted to individual researchers, and REU sites, which bring together larger numbers of students (usually from other campuses) under a common research theme. REU sites are expected to provide additional activities such as seminars on ethics, science communications, job strategies, and other professional development. Most REU participants present posters and/or talks at the end of their experience.

NSF requires MRSECs to have an REU site in which the majority of participants are from other campuses. REU programs often admit students from a range of degree programs, which provides a path to graduate materials science study for students with non-materials-science undergraduate degrees. Some MRSECs work with students from departmental-based REU sites. REU programs may be funded directly from the MRSEC budget or through a separate grant proposal to the REU program.

Some MRSECs provide other research opportunities for undergraduates. In addition to employing local undergraduates year-round, some programs bring undergraduates (with or without accompanying faculty members) from minority-serving institutions or primarily undergraduate institutions for summer research. Some programs offer the opportunity to continue research collaborations during the academic year as well.

Research Experiences for Teachers (RET) is an NSF-wide program offering K-12 teachers opportunities to work with a MRSEC during the summer (see Box 4.2). The RET program has as a goal involving teachers in research and transferring the knowledge gained from these experiences to the classroom. Teachers typically spend from 6 to 8 summer weeks with the MRSEC and receive a stipend of up to 2 academic months' salary. Some programs continue interactions with the teacher and his or her students during the academic year, which may include MRSEC researchers visiting schools or students visiting MRSEC laboratories. Many MRSECs provide a small amount ($1,000) of funding for supplies or other materials necessary to implementing curriculum. Some programs allow teachers to participate for more than 1 year, while others limit participation to 1 year. The implementation of the RET from MRSEC to MRSEC varies much more than that of the REU program. Some RET programs essentially duplicate the REU structure (and may have common activities). At the other extreme are programs that have little or no formal research component, with teachers developing materials-science-related curricula to use in their classrooms.

Some MRSECs have involved exceptional high-school students in research. These experiences range from a few weeks in the summer to year-round involvement. High-school students may participate in REU and/or RET activities, and some have made presentations at local and national meetings and have coauthored publications.

BOX 4.2
Research Experiences for Teachers

The Research Experiences for Teachers (RET) program was formalized by the National Science Foundation (NSF) Directorate for Engineering in FY 2001 with these goals: "to involve middle and high school teachers in engineering research in order to bring knowledge of engineering and technological innovation to the pre-college classroom."[1] Guidelines sent to the Materials Research Science and Engineering Centers (MRSECs) in January 2004 were based on a preceding "Dear Colleague Letter" of January 26, 1999, circulated by the NSF Directorate of Mathematical and Physical Sciences (MPS) and including the following directives (in fact, the first RET-type program was probably the Research Experiences for Science Teachers [REST] program at the Northwestern University MRSEC in 1999):

> The RET activity is designed to allow the participation of K-12 teachers in established Research Experience for Undergraduate (REU) sites. Eligible for this supplement are regular REU sites supported by MPS and all Centers that support REU site-like programs (such as MRSECs).
>
> The request should describe: 1) The plan for teacher activities and the nature of involvement with the REU site program; 2) Plans for incorporation of new learning into the K-12 classroom; 3) The teacher recruitment plan and the selection process; 4) The PI's [principal investigator's] experience in involving teachers or any previous collaborative work with teachers; 5) Plans for assessment of the program; and 6) Progress for any previously funded RET activity.
>
> Funding for the supplement may include up to two months of the teacher's annualized salary. As with all REU awards, indirect costs are not allowed, an administrative allowance limited to 25% of the teacher stipend is permitted. Requests may be for one year or for a 3-year period.

The RET program is further described by program solicitations originated in the NSF Directorate for Engineering and Directorate for Biological Sciences.

> Through these partnerships, the RET program aims to build long-term collaborative relationships between both in-service and pre-service K-12 teachers, community college faculty, and the engineering research community; support the active participation of these teachers and future teachers in research and education projects funded by NSF/ENG; facilitate professional development of K-12 teachers and community college faculty through strengthened partnerships between institutions of higher education and local school districts; and encourage researchers to build mutually rewarding partnerships with teachers. (NSF Program Solicitation 03-554)
>
> For example, the teacher may participate in the design of new experiments, modeling or analysis of experimental data or other activities that will result in intellectual contributions to the project. Since it is expected that the RET supplement experience will also lead to transfer of new knowledge to classroom activities, the RET supplement description should also indicate what sustained follow-up would be provided to help in translating the teacher's research experience into classroom practice. (NSF Program Solicitation 05-524)

---

[1]National Science Foundation, *Program Solicitation for Research Experiences for Teachers: Supplements and Sites*, NSF 03-554, Washington, D.C., 2003.

Perhaps the most direct education impact of MRSECs is on the graduate students who research and learn within the program. These students are exposed to multiple principal investigators and shared facilities, and they often participate in center-based journal clubs and discussion groups. From its site visits, the committee learned that in some cases MRSECs are the great enabler of this broadened educational experience, and that in other cases MRSECs are the result of a preexisting disposition on the campus. However, independent of whether MRSECs uniquely train graduate students, this is an area of significant value for the program.

Goal: Improving Understanding and Appreciation of Science and Engineering

In addition to encouraging young people to pursue science and engineering study, some MRSEC EO programs attempt to increase the scientific literacy of the current and future citizenry. These efforts include approaches that are both formal (K-12 schools and universities) and informal (talks for the public, educational sessions for legislators or reporters). Programs include improving content knowledge by means of involvement in local K-12 and college-level education, developing curricular materials, and informing policy makers. Other approaches focus on improving understanding of how research works, contributing to awareness of career options, and promoting general enthusiasm for science and engineering.

At the college level, MRSECs have developed courses and curricula for graduate and undergraduate courses. Most of these classes are highly interdisciplinary, focus specifically on the MRSEC topic area, and are designed to involve students in different departments. Some are cotaught by faculty members from different disciplines. Some MRSECs report developing and/or implementing new pedagogical techniques that enhance student learning (i.e., active learning techniques). See Appendix D, "Further Information on Education and Outreach Activities."

A broad range of activities at the K-12 level includes curriculum development, classroom visits from MRSEC researchers, professional development activities for teachers, summer enrichment programs for teachers and/or students, and laboratory visits. Because of the standardized testing requirements imposed by the No Child Left Behind Act of 2001 (Public Law 107-110), many K-12 activities focus on general science and/or engineering rather than on the research theme of the MRSEC.

Outreach to the public generally occurs in informal settings including lectures, demonstration shows, building or contributing to exhibits at science museums, and workshops for policy makers, journalists, and business people. MRSECs also hold open houses, sponsor science days for parents and children, and may develop audio and/or video materials.

Goal: Broadening Participation

One of the few activities specifically mandated by NSF is that of increasing the participation of women and other underrepresented groups in MRSECs. Although broadening participation by people in these groups has always been an important part of the Broader Impacts requirement of NSF proposals, MRSECs have been required to develop formal "diversity plans" since 2001 and are expected to show results from those plans over the course of the MRSEC grant.

Table 4.1, which displays demographics of MRSEC participation, shows that MRSECs are having the most success at broadening participation in undergraduate and pre-college audiences but that the involvement of underrepresented minorities in particular needs to be much improved at the graduate-student and higher levels. For reference, Table 4.2 shows the overall percentages of women and minorities involved in materials research and related fields. Some of the strategies that MRSECs use to broaden participation include partnerships with minority-serving institutions (some through the PREM program—see Box 4.3) and/or women's colleges; interactions with K-12 schools serving underrepresented populations; alliances with professional associations for minority scientists and engineers; and participating in or holding special programs for underrepresented groups.

MRSEC–MRSEC Interactions

One of NSF's goals for the MRSEC program is for it to be a network of centers focused on advancing research and education in materials science and engineering (MSE). Some aspects of the EO program are shared by many MRSECs, which offers opportunities to share information and resources. Some of these interactions have developed around common programs such as REU and RET, while other efforts (such as the participation of EO coordinators in education-themed MRSEC

TABLE 4.2 Percentage of Women and Underrepresented Minorities (URMs) in the Fields Most Represented in Materials Research Science and Engineering Centers

	Physics		Chemistry		MS&E	
	Women (% in field)	URM (% in field)	Women (% in field)	URM (% in field)	Women (% in field)	URM (% in field)
Bachelor's degree	23	10	50	16	30	7.5
Master's degree	21	7	46	8	27	5.8
Ph.D.	14	6	35	6	16	6.0

NOTE: Figures for bachelor's and master's degrees are for 2002, while figures for Ph.D. degrees are from 2003. Figures do not include "unknown" designations. MS&E, Materials Science and Engineering.
SOURCE: National Science Board, *Science and Engineering Indicators 2006*, National Science Foundation, Arlington, Va., 2006. See http://www.nsf.gov/statistics/seind06/.

BOX 4.3
Partnership for Research and Education in Materials

The Partnership for Research and Education in Materials (PREM) was established in the NSF's Division of Materials Research (DMR) in 2004 to develop materials research and education partnerships between minority-serving institutions (MSIs) and MRSECs. The 10 currently active PREM awards are listed below, with their dates of initial award in parentheses:

- California Institute of Technology–California State University at Los Angeles (2004)
- Carnegie Mellon University–Florida Agriculture and Mechanical University (2004)
- University of Pennsylvania–University of Puerto Rico at Humacao (2004)
- University of Wisconsin–University of Puerto Rico at Mayaguez (2004)
- Princeton University–California State University at Northridge (2006)
- University of California at Santa Barbara–Jackson State University (2006)
- Cornell University–Norfolk State University (2006)
- Johns Hopkins University–Howard University (2006)
- Cornell University–Tuskegee University (2006)
- Harvard University–University of New Mexico (2006)

PREM strives to create cooperative research teams and provide experimental facilities to the partner institutions, thus providing additional research and education opportunities for students and faculty. Since PREM is still a rather new program, it is too early to determine its impact on this problematic issue.

directors' meetings) have been initiated by the NSF. These interactions are summarized in Appendix D.

Distribution of Education and Outreach Resources

MRSECs spend approximately 10 percent of their budgets on EO; however, this figure may be misleading, because some activities are funded by supplemental grants. Some MRSECs fund their REU and/or RET activities entirely from the MRSEC budget, while others fund them from a separate grant or supplement. The origin of the funding and whether it is being accounted for in the annual reports is not always clear. Therefore, a representative group of MRSECs was asked to provide more detailed information about how their EO budgets are distributed. Funding for the RET program comes entirely from the Office of Multidisciplinary Activities in the NSF Directorate for Mathematics and Physical Sciences, regardless of whether the RET program is included in the original MRSEC budget, is a separate grant, or is a supplement to the MRSEC budget. The analysis of the detailed budget breakdowns shows:

- The majority (more than 75 percent for most MRSECs) of the EO budget goes to research-related EO (involving students from high school to graduate school and teachers in research, research-related conferences and workshops, and the formation of research-based partnerships with primarily undergraduate and/or minority-serving institutions) and to EO personnel costs.

- The majority of the K-12 and public outreach programs, with a very few exceptions, comprise a very small fraction of the MRSEC EO budget. The MRSEC contributes a few percent or less of the total funding for the majority of these activities.
- Many MRSEC EO activities receive funding from sources outside the MRSEC grant per se. In addition to REU and/or RET supplements, institutions may provide support, and a number of MRSECs lead or participate in separate education grants, such as the Integrative Graduate Education and Research Traineeship, PREM (see Box 4.3), Nanoscale Undergraduate Education, and Graduate Fellows in K-12 Education.

IMPACT OF MRSEC EDUCATION AND OUTREACH PROGRAMS

EO plays an important role in supporting U.S. excellence in science, technology, engineering, and mathematics (STEM), and the MRSEC program invests significant resources in EO. The study committee addressed two questions: (1) Are MRSECs meeting NSF's and their own self-determined goals in education and outreach? (2) Are those goals the best use of MRSEC resources? MRSECs were asked to provide the committee with copies of any evaluation instruments and/or studies they had conducted on their EO programs.

Issues Affecting the Evaluation of MRSEC Education and Outreach Programs

The committee received evaluation information from 13 MRSECs. Some of these evaluations were of separately funded programs. Of the remainder, the majority of the evaluations was of REU and RET programs. These evaluations focused primarily on logistics and participant satisfaction with the program. From these evaluations and the data described previously, the committee observed that:

- EO programs span a broad range of programs that serve many different audiences and, with the possible exceptions of REU and RET, are specialized to fit local situations. While this range of activities is encouraged by the NSF, each MRSEC has to manage multiple, often very different activities.
- Many MRSEC EO activities are leveraged by other programs, making it difficult to identify what can be attributed to the MRSEC.
- The data available are not sufficient for thoroughly evaluating MRSEC EO impact. The evaluations received by the committee rely almost entirely

on self-reporting during the program, which lacks the objectivity and perspective necessary for a meaningful evaluation. Self-reporting provides information on the participant's perception of value, but it provides no evidence of efficacy. Evaluations of this type rarely collect the kind of data necessary to compare the outcome of an activity with alternative activities that have the same goals.

Are MRSECs Meeting Their Own and the Program's Goals?

The annual reports that NSF requires from the MRSECs include information on EO activities, and these internally generated documents provided additional information about whether MRSECs are meeting their self-determined goals. Individual MRSEC goals are consistent with the NSF goals described in the MRSEC program solicitation. The available data indicate that MRSECs generally are meeting their (and NSF's) goals; however, much of the evidence supporting this statement is anecdotal and self-reported. Additional objective evidence would greatly strengthen this conclusion. Research-related activities—especially the REU and RET programs—were evaluated with much greater frequency by the MRSECs that provided the committee with evaluation information than were other types of activities. This may be due to the availability of evaluation instruments generated by members of these communities. Heavily leveraged activities may be evaluated more thoroughly; however, the evaluation is likely to originate from (and be funded by) the primary grant supporting the activity.

According to the information provided to the committee, the formative evaluations submitted most frequently use open-response and/or multiple-choice surveys to probe participant opinions. A much smaller number of MRSECs use observation of participant behavior, journal analysis, and formal or informal interviews. The summative evaluations submitted also rely primarily on participant surveys, with a much smaller number of MRSECs using content tests, classroom observations, and participants' journals. MRSEC evaluations tend to be weighted toward the summative; however, few annual reports addressed how their respective programs responded to formative or summative evaluation results. Few data were available relating to whether MRSECs formally adjusted their goals over the course of the grant or indicating the role that evaluation played in determining goals for future proposals. The EO portions of the annual reports tend to focus on what happened and on who participated rather than on outcomes.[1]

[1]National attention on evaluating the longitudinal impact and effectiveness of education programs in science, technology, engineering, and medicine (STEM) has been growing. A review of STEM education programs across the federal government finds that few programs have been rigorously evaluated and little is known about their impact on students. This report, by the Academic Com-

REU/RET

Formative evaluations of MRSEC REU and RET programs are designed primarily to identify situations requiring intervention. Summative evaluations assess participant impact and evaluate program structure, with most items focusing on program organization (seminars, social activities, workshops), logistics (dorms and travel), the nature of the research project, and the experience with the mentor. Evaluations investigating impact and outcomes focus primarily on the following:

- The participants' perception toward science and research,
- The participants' confidence in science and in doing research,
- Knowledge and skills gained from the program, and
- Career plans.

For RET programs, surveys also asked about the following:

- Participants' attitudes toward teaching science, and
- Plans to integrate what they have learned into their teaching.

Selected results from MRSEC evaluations of the REU and RET programs are shown below. According to the responses reported by present MRSECs in response to the committee's survey, nearly two dozen MRSECs have supported RET-type programs.

- Almost all REU and RET participants express high satisfaction with their experience.
- REU participants report that their experiences impact their career choices. Centers that track their participants past the end of the program report that a high rate of students who attend MRSEC REUs pursue graduate study in MSE; however, since REU participants are self-selected and a comparison group is rarely involved, it is difficult to attribute this directly to the REU program.
- Participants reported gains in skills, including communication skills, specific science content, self-confidence, and understanding of scientific research methods. A few MRSECs asked mentors to provide independent evaluations of progress in these areas.

---

petitiveness Council, recommends that funding for federal programs to improve STEM education outcomes "should not increase unless a plan for rigorous, independent evaluation is in place." See Department of Education, *Report of the Academic Competitiveness Council*, Washington, D.C.: Government Printing Office, 2007, p. 3.

- The primary complaints of participants focused on mentors who were unprepared, did not clearly communicate goals and expectations, were unavailable, or chose projects that could not be completed in the available time.
- Most MRSECs report that RET participants plan to integrate what they learned into their classroom practice; however, there has been limited follow-up as to the extent to which this happens.
- RET participants enjoyed seeing the connections between what they did and their own classroom curricula.
- Mentors (faculty, postdoctoral researchers, and graduate students) generally view their involvement in REU and RET programs as being rewarding and as providing a service to the community.
- Graduate student and postdoctoral mentors believe that the experience was valuable preparation for their future professional responsibilities. REU mentors reported personal and professional benefits, including developing skills in planning and executing a short research project, learning how to manage time effectively, and experiencing the satisfaction of seeing a student mature and become proficient in scientific research, although sometimes at the cost of slower research.

The available evaluations show a high level of participant satisfaction; however, there are limitations on the conclusions that can be drawn. In addition to the inherent limitations of self-reported evaluations, few MRSECs follow up with participants after the program to determine whether they had acted on intentions expressed during or immediately after the program. Many of the items surveyed, such as perceived self-confidence, are difficult to measure objectively. Finally, because most of the programs are self-selecting and few attempt to establish a comparison group, it is difficult to determine what impact can be attributed specifically to the MRSEC experience.

The committee believes that the available evidence shows that MRSECs are doing an excellent job of meeting the goals set for the REU program by providing an environment conducive to its goals. The REU program is one of the areas in which MRSECs have succeeded in attracting diverse students. There appear to be some discrepancies between the goals that individual MRSECs have for their RET programs and those expressed by NSF for the program. One difficulty is that there is no formal RFP for the RET program as there is for the REU program; this situation has produced some confusion about what the goals of the RET program actually are. In evaluating the information provided to the MRSECs about RET, the committee is concerned that some MRSECs are not meeting the goals that NSF has for the program. Some implementations of the RET program focus primarily on curriculum development, with research playing a very limited role, if any. Although the RET program succeeds in involving women and underrepresented minorities, the committee is very concerned that there are Research Experiences for Teachers programs that do not focus on research.

K-12 and the Public

Few of the programs for K-12 and the public are evaluated at the same level at which the REU and RET programs are evaluated, so it is difficult to evaluate whether goals for these programs are being met. Well-evaluated programs often conduct those evaluations and reviews under the auspices of supplemental funding from other sources. The committee saw many examples of innovative programs that were enthusiastically received and executed; however, impacts of these programs beyond generating enthusiasm cannot be determined.

MRSECs EO programs for K-12 and the public are highly responsive to local needs and interests. Many programs are driven by individual researchers who donate their time and effort. In many cases, researcher participation is facilitated by the EO coordinator, who handles logistics and organization. The ability to address local needs is a positive outcome of the flexibility allowed by the MRSEC program.

Broadening Participation

Although the MRSEC program as a whole is making strides in increasing the involvement of women at all levels, there is considerable variation among MRSECs. Few MRSECs attract sizable numbers of underrepresented minorities, in part because of the overall small numbers and the competition among institutions. The PREM program (see Box 4.3) is too new to evaluate, but long-term programs such as this have much higher potential for impact than is possible for isolated activities such as "Women in Science" days. The shifting national demographics demand that the materials science and engineering community increase efforts to broaden participation. There has been no attempt at a MRSEC-wide effort in this area, but such a strategy may be worth pursuing.

Evaluating the Appropriateness of the Goals

The second part of the committee's task was to evaluate whether the EO goals are appropriate. The impact, or potential impact, of the programs was the most important consideration, with a second consideration being whether there were alternative programs with similar goals that might be more efficacious. Finally, the committee evaluated the programs to determine whether the MRSEC provided any unique aspects that would not be duplicated by the same program run outside the MRSEC.

REU

Involving undergraduates in research continues to be an integral part of the NSF portfolio.[2] The widespread involvement of undergraduates in research has generated a significant research base that addresses the impact of undergraduate research experiences (including, but not limited to REU).[3-5] This research concludes:

- Undergraduate research experiences help students clarify their career goals, including whether they want to continue STEM study, the specific type of subdiscipline in which they choose to continue, and what graduate school they will attend;
- Undergraduate research experiences provide an apprenticeship in which students learn to "think and work like scientists" alongside working scientists. In particular, students appreciate how science is done, gaining a perspective often ignored in textbooks, and they learn to work independently and to rely on their own judgment;
- Students learn specific technical skills; and
- Most students experience personal gains, including increased confidence in their ability to be successful in STEM fields.

There is ample evidence that involving undergraduates in research is positive and has a great impact on the participants, including the mentors. Researchers are overwhelmingly positive about the program and their participation as mentors. Because this type of activity is so widespread, there are a number of assessment

[2]REU evaluation instruments are available from the MRSEC Web site (http://www.mrsec.org/links/).

[3]See A.-B. Hunter, S. Laursen, and E. Seymour, "Becoming a Scientist: The Role of Undergraduate Research in Students' Cognitive Personal and Professional Development," *Science Education*, 91 (1): 36-74, January 2007.

[4]See Susan H. Russell, "Evaluation of NSF Support for Undergraduate Research Opportunities," *SRI International*, May 2006, Menlo Park, Calif.: SRI International, 2006. See http://www.sri.com:8000/policy/csted/reports/university/documents/URO%20FollowupSurveyReport%20for%20WebApr%2028%2006.pdf.

[5]See E. Seymour, A.-B. Hunter, S. Laursen, and T. DeAntoni, "Establishing the Benefits of Research Experiences for Undergraduates in the Sciences: First Findings from a Three-Year Study," *Science Education* 88: 493-534, 2004.

tools, results, and best practices that are shared at disciplinary and REU-specific conferences.

REUs are especially appropriate for MRSECs because they offer undergraduates unique experiences owing to the interdisciplinary environment. REU programs in materials science are especially valuable to students at institutions without undergraduate materials science programs, as they open the door to graduate MSE study. The REU program is one of the areas in which MRSECs are attracting a diverse group of students, making it an important component in building the scientific and technical workforce.

RET

The RET program is newer than the REU program, so there is commensurately less information about its impact.[6] Some preliminary conclusions can be drawn from the published literature (which comes from MRSEC and non-MRSEC RET programs),[7-9] with the caveat that the studies are limited in number and scope.

- The primary impact of research experience on teachers is in improving their understanding of how science is done, their knowledge of current science, their awareness of the types of people who are and who become scientists, their awareness of STEM career opportunities, and in increasing their willingness to take on leadership roles.
- Constraints on teachers (time, standardized testing, and student capability) make it difficult to bring content from their research into the classroom. The majority of teachers focus on translating their understanding of scientific *process* rather than specific content to their students.
- Teachers who have a research experience exhibit an increased use of inquiry-based and problem-solving techniques with students, heightened emphasis on scientific process (working in groups, using graphs and charts), expect more students to design their own experiments, have more

[6]RET Network Web site, http://www.retnetwork.org/evaluation.htm.

[7]Carol S.C. Johnston, *Translating the RET Experience to the Classroom*, Redwood City, Calif.: Conference on Assessing, Determining, and Measuring the Impacts of the Research, 2003.

[8]J. Dubner, S.C. Silverstein, N. Carey, J. Frechtling, T. Busch-Johnsen, J. Han, G. Ordway, N. Hutchinson, J. Lanza, J. Winter, J. Miller, P. Ohme, J. Rayford, K. Sloane-Weisbaum, K. Storm, and E. Zounar, "Evaluating Science Research Experience for Teachers Programs and Their Effects on Student Interest and Academic Performance: A Preliminary Report of an Ongoing Collaborative Study by Eight Programs," *Journal of Materials Education* 23: 57-69 (2001).

[9]Kevin Dilley, "How Do You Measure RET Success?" Redwood City, Calif.: Conference on Assessing, Determining, and Measuring the Impacts of the Research, 2003.

students involved in science fair projects and science clubs, and talk more to their students about STEM careers.
- The most comprehensive of the published studies shows an increase in students' content knowledge with such teachers as measured against comparison classes on standardized tests; however, few studies have addressed this important impact.
- Many programs report that gains (regardless of type) come only after sustained teacher participation, which may include multiple summer RET experiences or a program that continues throughout the school year. Changes in teaching practice and/or student content knowledge may also take a year or two after the RET experience to be evident.

Although the preliminary results indicate that RET has potential for high impact on student learning and future professional interests, the committee has two concerns about its role in the MRSEC program. The literature indicates that the most-transferred elements of the teacher research experience are the process skills derived from actually doing research. A number of MRSEC programs focus entirely or primarily on curriculum development, without a significant research component. The RET is not intended to be a curriculum-development program. NSF supports curriculum development through separate programs that require peer-reviewed proposals with formal evaluation and dissemination plans. In view of the emphasis on standardized testing in K-12, the committee is concerned that RET-based curriculum-development programs may have very limited impact.

A second concern is the lack of evidence as to how involving teachers in research ultimately affects their students. Although the preliminary data suggest that increased student learning or even improved attitudes toward mathematics and science should result, the majority of MRSEC evaluations focus on logistics and self-reported satisfaction level. It is important to establish the impact of the MRSEC RET and especially whether unique outcomes result from an RET in a MRSEC compared with outcomes in other research fields. It is impossible to judge the value of the RET program within the MRSEC portfolio without an accurate representation of the benefits. The resources currently invested in the RET program might have more impact if focused on other types of professional development activities for teachers.

K-12 and the Public

The range of education and outreach programs targeted to K-12 students and the public is extremely broad. With a few notable exceptions, the evaluation of these programs is minimal, making it impossible to judge the efficacy of each program. Many of the programs for these audiences are leveraged heavily by other funding sources, making it difficult to determine whether the MRSEC involvement has any impact.

There are many convincing arguments for why MRSECs should be involved in K-12 and public outreach. Getting children interested in science early and maintaining that interest is critical to producing future scientists and a scientifically literate citizenry. Many students never hear about "materials science and engineering" in K-12, which may decrease the likelihood of their pursuing MSE study in college or even of appreciating the contributions that materials science and engineering make to their quality of life. Most programs targeted to K-12 students and the public are highly responsive to local needs, which is important; however, some MRSEC participants felt that they were downgraded in reviews for not consistently producing new and innovative programs rather than continuing to execute a program that they know works and fulfills a recognized need. Most MRSEC participants say that they enjoy participating in EO, and researcher enthusiasm is a large driving force.

The difficulty in endorsing these EO programs is the lack of evidence as to their impact relative to the time and effort required to run them. It is the committee's impression that the broad range and large numbers of programs in this category reflect the pressure that MRSECs feel to address every possible audience. Regardless of the origin of this pressure, the result appears to be a type of EO "arms race": each MRSEC feels compelled to outdo the others by being able to cite a broad range of programs that reach large numbers of people of all ages. The unfortunate result is an emphasis on quantity over quality. There are a few exemplary programs in this category; however, executing a large number of programs with limited impact is not as effective as implementing a smaller number of high-quality programs that have the budget and responsibility for meaningful evaluation.

Preparing Future Researchers for Participation in Education and Outreach

An important and potentially overlooked aspect of the MRSEC EO program is that the involvement of graduate students, undergraduate students, and postdoctoral researchers in EO programs helps prepare them for future roles as materials science researchers and educators. The broad range of activities gives them myriad opportunities for participating. EO programs help researchers learn effective ways to engage students and the public while reinforcing the importance of integrating research and education. This is especially important for graduate students and postdoctoral associates, from whom these activities will be expected in the future. While it would be interesting to investigate how the MRSEC research atmosphere influences students at these levels, data from NSF were somewhat limited. NSF was able to provide data on decisions of MRSEC Ph.D.s to pursue careers in industry (see Figure 5.3 in the next chapter), which when compared with MSE overall showed little difference in outcome. The committee, unfortunately, was not able to analyze Ph.D. student choices of academia and other arenas, as well as postdoctoral choices.

FINDINGS AND RECOMMENDATIONS

Education and outreach have an important role in developing the scientific and technical workforce, in educating the public about scientific issues, and in broadening the participation of women and other underrepresented groups. MRSECs have a great opportunity to contribute to this mission through their EO programs.

Conclusion: The MRSEC education and outreach program has impacts on the NSF mission to educate and prepare the nation's future workforce.

- MRSECs provide unique opportunities for interdisciplinary research experiences that are different from those an individual student would experience in a single-investigator laboratory.
- MRSECs foster environments that support interactions with other programs to leverage funds and coordinate activities across campuses and disciplines. This culture leaves a vital imprint on students who work in MRSECs.
- MRSECs foster a mind-set of outreach and a sense of responsibility in current and future researchers.
- The centralized EO infrastructure that a MRSEC offers empowers researchers to engage in EO who would not have ordinarily done so.

The MRSEC EO requirement facilitates the involvement of interested researchers at all levels. EO coordinators are valuable participants who develop programs, arrange logistics, and build the partnerships that make it possible for researchers to be effectively involved in EO. The MRSEC EO requirement allows faculty members to pursue their EO interests and can provide funding and infrastructure support for that pursuit.

General Finding: The most significant and well-documented contribution of MRSEC EO programs is the preparation of future researchers at all levels.

Research-related education and outreach activities leverage MRSEC strengths and expertise. MRSECs can provide unique opportunities for interdisciplinary research experiences that are different from those that an individual would experience in a single-investigator laboratory. Although broadening the participation by women and underrepresented groups remains a challenge, the greatest contributions to meeting this challenge often come from EO programs such as REU and RET.

Conclusion: Although the committee's impression is that most MRSECs are doing good to excellent jobs with their EO programs and that many of these programs have a significant impact on their audiences, the lack of data to support these assertions poses a serious problem for NSF as it seeks to make the most efficient use of its resources.

NSF manages the MRSEC program from a scientific and engineering research perspective. It is nonprescriptive, with few defined limits or requirements. The lack of specificity regarding EO expectations has led to some innovative, potentially high-impact programs; however, this lack of specificity also has led many MRSECs to try to carry out some type of activity in every aspect of EO that they see their peer (competitor) centers doing.

REU and RET programs are much more likely to be evaluated (in general and especially by the MRSECs), although the evaluations focus primarily on logistics and self-reported participant perceptions. The quality of evaluations on other EO components varies greatly. MRSECs are reviewed primarily on the breadth of activities and the number of participants and not on documented outcomes.

General Finding: The future impact of MRSEC EO activities is threatened. The continued lack of specificity in EO expectations at the agency level has led to an emphasis on quantity over quality and innovation over impact.

It is evident to the committee that there is a multiplicity of EO activities in the MRSEC program and that the lack of guidance from NSF to the MRSECs and reviewers has contributed to what appears to have become a less productive enterprise. This has produced an emphasis on quantity over quality and on doing something new for its own sake rather than choosing to implement proven strategies.

General Finding: Most MRSECs feel compelled to participate in many disparate education and outreach activities. This approach often does not make optimal use of the MRSECs' strengths, dilutes their potential impact, and in fact reduces the likelihood of determining what that impact is.

There is a perception that the demands of the EO program have grown significantly since the original inception of the MRSEC program. While the solicitations for the program show most growth in demands, the broad portfolio of activities, even in the smallest MRSECs, suggests that MRSEC resources are being spread too thinly and that the impact of those resources is being diminished.

This perception should not be taken to suggest that the community does not value EO. Though the tight coupling of resources to support EO programs makes it difficult for the committee to draw explicit conclusions about the appropriateness of the level of researcher involvement, the overwhelming majority of MRSEC participants expressed a belief that EO is important and indicated that they enthusiastically participate in EO activities. Nevertheless, there is a strong belief among the MRSEC participants, present and prospective, that the selection process rewards quantity over quality and innovation over impact. Two examples were most often mentioned by these individuals:

- The belief that a MRSEC must reach all audiences, including K-12, undergraduate and graduate students, and the public; and
- The belief that continuing an existing, successful program is received less favorably than proposing something new.

The emphasis on breadth has led to evaluations that consist primarily of counting numbers of attendees, because the programs are so diffuse that more meaningful evaluation is impossible without funding from other sources. Some programs focus on generic outreach that has little to do with the MRSEC focus, much less materials science and engineering. While this type of outreach is important, it does not leverage MRSEC resources.

Existing MRSECs mentioned that renewal reviews value doing something new over continuing programs that have been shown to be effective. The larger question is whether MRSECs should be required to innovate in the EO component of their programs or whether the focus should be on using best practices to make an impact on their communities.

Focusing MRSEC resources into a select number of programs that address the local strengths and needs makes much more sense than trying to reach all audiences. The MRSECs that are successful in reaching a variety of audiences often are those with significant external funding for EO.

Recommendation: Education and outreach should continue to be part of the overall MRSEC portfolio; however, MRSECs should focus resources on programs with proven high impact that leverage each MRSEC's unique research strengths and that can be meaningfully evaluated.

The committee believes that EO is an important part of the MRSEC program but that steps can be taken to increase its effectiveness. In particular:

- MRSECs should focus on a limited number of activities that are aligned with MRSEC research goals, are consistent with the MRSEC size, leverage participant expertise and interest, and address local needs.
- Because of their documented impact, REU programs should continue to

be required; providing research opportunities for faculty and students at predominantly undergraduate and minority-serving institutions should be strongly encouraged.

- MRSECs that offer RETs should provide teachers with research experiences in materials science and engineering. The RET is not meant to be primarily a curriculum-development program.
- Other EO projects should be peer reviewed by materials research education experts during the MRSEC proposal/review process. The best of these projects should be funded as long as the overall MRSEC is funded.

MRSECs, especially those with smaller budgets, are trying to do too much with the resources they have. This is not intended to discourage MRSECs from developing and executing EO activities; however, resources would be better directed by funding a smaller number of high-quality, research-oriented activities whose impact can be meaningfully determined.

There is ample evidence that the REU program has highly desirable impacts, and MRSEC researchers generally are enthusiastic and committed about their participation in REUs. MRSECs offer unique opportunities for students to get involved in interdisciplinary research at early stages of their careers and are an important pathway to graduate study in materials science and engineering.

The RET recommendation is tempered by the committee's concern that the impact of the RET program is largely undocumented. The RET program is NSF-wide, so the lack of data is not solely a MRSEC issue. Cooperative efforts to document the impact of the program, as have been done with the REU program, are necessary. However, validating the program is beyond the scope of what should be expected as part of a MRSEC EO component. Further, MRSEC RETs that do not focus primarily on providing research experiences for teachers are not addressing the intention of the RET program. All RET programs should focus on research.

MRSECs should be encouraged to form partnerships with predominantly undergraduate and minority-serving institutions, and to extend research opportunities to faculty and students from those institutions. Participation in the PREM program has been and should continue to be encouraged. These activities are especially important in increasing the diversity of materials science and engineering.

One way to accomplish this is by having MRSECs' EO projects beyond the research-related activities discussed above evaluated separately by materials research education experts, as available. The committee believes that education expertise is more valuable than materials research expertise when evaluating these activities. Program managers would then fund the highest-ranked projects from those proposed by successful MRSECs.

Recommendation: In the context of the above recommendation, NSF should develop and support the MRSEC education and outreach community in sharing and facilitating ideas and resources, including best practices, for all activities. This would be especially helpful in the area of increasing the participation of underrepresented minorities.

The collective impact of MRSECs in education and outreach could be enhanced by increased cooperation and coordination among the centers. Progress is being made in this direction, but more is possible. Despite the broad range of research interests, all MRSECs have common EO goals and activities, and an overall shift in emphasis from innovation to impact would make it easier for MRSECs to share best practices. This would facilitate the distribution of EO materials already developed and would decrease local reinvention of existing EO materials. In this vein, MRSECs should adopt a standardized evaluation instrument used at all sites to ensure that programs are using the established best practices. MRSECs should be encouraged to add to that evaluation; however, the adoption of a standard evaluation establishes a baseline for acceptable performance.

The National Research Center Educator's Network could be a starting point for this community; however, the meetings of EO coordinators run by NSF have the advantage of being run simultaneously with MRSEC directors' meetings, which keeps directors informed about EO issues. EO coordinator meetings should be held annually, and the NSF MRSECs should establish an EO coordinators' executive committee (similar to that of the directors) to facilitate coordination, communication, and dissemination. This group should plan the workshops (with input from the members) to address long-range strategic issues and to provide continuity.

The PREM program is an excellent example of how NSF can act as a catalyst for activities that involve women and underrepresented minorities in materials science and engineering research. The committee believes that centralized activities such as PREM have a much higher probability for success than does leaving each MRSEC to its own resources. NSF should leverage the experience of its MRSECs to identify and share successful strategies in this area, not just with other MRSECs, but with the materials science and engineering community as a whole.

Recommendation: NSF should provide appropriate guidance to MRSEC applicants and reviewers in order to refocus education and oureach activities and ensure the program's effectiveness.

It is evident to the committee that there is a multiplicity of EO activities in the MRSEC program and that the lack of guidance from NSF to the MRSECs and reviewers has contributed to what appears to have become a suboptimal enterprise. This should not be so. Reviewers should receive clear instructions about the role of EO in the MRSEC: the impact of a MRSEC's EO program should be of cardinal importance. Further, MRSEC EO programs have different objectives and therefore should not be evaluated using the same standards as those for research. NSF funds educational research under other programs, and major initiatives should be supported through those programs, with a separate review system.

5

Assessment of the Impact of MRSEC Collaboration with Industry

Throughout the history of the Materials Research Science and Engineering Centers program (MRSEC program), one important goal has been to promote "active cooperation with industry to stimulate and facilitate knowledge transfer among the participants and strengthen the links between university-based research and its application," as stated in the National Science Foundation (NSF) request for proposals. To implement this goal, each Materials Research Science and Engineering Center (MRSEC) is required to implement and execute a program for knowledge transfer to industry. The requirement is illustrated by MRSEC program solicitations issued since 1993 (e.g., NSF 93-106, NSF 95-89, NSF 97-98, NSF 99-125, etc.); representative excerpts are given below:

> [MRSECs] are expected to have strong links to industry and other sectors, as appropriate, and to contribute to the development of a national network of university-based centers in materials research. (MRSEC Program Solicitation NSF 99-125)

> [MRSECs foster] active cooperation among university-based researchers and those concerned with the application of materials research in industry and elsewhere. (MRSEC Program Solicitation NSF 99-125)

> MRSECs incorporate most or all of the following activities to an extent consistent with the size and vision of the center. . . . Active cooperation with industry to stimulate and facilitate knowledge transfer among the participants and strengthen the links between university-based research and its application. . . . (MRSEC Program Solicitation NSF 99-125)

Modalities of the industry cooperation are cited more explicitly in, for example, the following solicitation:

> Active cooperation with industry, to stimulate and facilitate knowledge transfer among the participants and strengthen the links between university-based research and its application. . . . Cooperative activities may include, but are not limited to: joint research programs; affiliate programs; joint development and use of shared facilities; visiting scientist programs; joint educational ventures; joint seminar series, colloquia or workshops; stimulation of new business ventures; involvement of external advisory groups; and industrial outreach programs. (MRSEC Program Solicitation NSF 97-98)

The MRSEC program stresses *flexibility* in each center's approach to setting research directions, seed projects, and outreach and education:

> Each MRSEC has the responsibility to manage and evaluate its own operation with respect to program administration, planning, content and direction. NSF support is intended to promote optimal use of university resources and capabilities, and to provide *maximum flexibility* . . . in developing cooperative activities with other organizations and sectors. . . . [Emphasis added.]

Thus, the National Science Foundation solicitations cited are consistent with the view that industrial collaboration in the context of the MRSEC charter should be an integral part of the MRSEC program. Its implementation should be flexible and consistent with the size, capabilities, mission, and vision of each individual MRSEC. It is important to note that industrial collaboration includes cooperation and interaction with relevant sectors involved with the application of materials research beyond just commercial industries. Consequently, industrial collaboration includes national laboratories and other federal entities (e.g., Department of Defense [DOD] laboratories) that apply the results of basic materials research to address important technical needs.

Materials science and engineering (MSE) is a key national resource. The recent decline in basic and exploratory materials research and development (R&D) in industry transfers the responsibility to universities not only to do transformational research in the area but also to transfer the knowledge obtained to industry for its application. Knowledge transfer to industry to facilitate the application of university research is especially critical for maintaining the preeminence of the United States in materials science and technology in today's global and technology-based economy. The effective transfer of knowledge from the university to industry is crucial to achieving the goals of the "American Competitiveness Initiative" promulgated by the President and Congress in early 2006. As such, it is most appropriate to continue industry collaboration and knowledge transfer as an integral part of the MRSEC program.

CURRENT INDUSTRIAL COLLABORATION AND KNOWLEDGE-TRANSFER ACTIVITIES

The initial step in evaluating the effectiveness and impact of industrial collaboration and knowledge transfer activities across the MRSEC program was to understand the range of current efforts. The committee's assessment of the current situation, which considered the efforts and results in this area over the past several years, was based on numerous teleconferences with MRSEC directors, discussions with industry participants, site visits to MRSECs, study of the MRSEC annual reports, and consideration of written responses from MRSECs to questions from the committee. An especially valuable perspective was provided by the November 2004 report of the MRSEC Directors Industry Working Group—the Working Group was chaired at the time by Michael Ward—which documented much of the ongoing industrial collaboration activities for 2002-2004 period. The "Ward report" did an excellent job of meeting its stated purpose, which was to "evaluate industrial participation with the MRSEC program as a whole,"[1] rather than focusing on specific activities or best practices at individual centers.

The information gathered by the committee provided a self-consistent picture of the industrial collaboration effort, which was in line with the NSF view that there are numerous effective ways to address the program goals for industrial collaboration and knowledge transfer. Workshops, short courses, and symposia were among the most common approaches to engaging industry and disseminating knowledge. Most of these meetings focused on specific technical topics. As documented in the Ward report, MRSECs fully or partly sponsored 22 such meetings in 2002 (916 total industrial attendees), 40 in 2003 (1,541 industrial attendees), and 43 in 2004 (1,620 industrial attendees).[2] MRSEC annual reports from 2005 suggest that the number of workshops, short courses, and symposia has remained at a similar level. The obvious advantage to this type of activity is the ability to promote broad engagement among MRSEC students and faculty and interested industrial representatives. It was emphasized by several MRSEC directors that student participation in these meetings was very important for enabling them to interact with industrial scientists and managers. These interactions were especially important at campuses that do not have a strong tradition of industrial engagement. From an industry perspective, the breadth provided through a MRSEC-sponsored technical event was seen as a value-added way of engaging a broad faculty group.

Many of the technical meetings sponsored by MRSECs are advertised, such as

[1]M. Ward, *MRSEC Industry Outreach/Education Activities Survey*, unpublished, November 22, 2004.

[2]M. Ward, *MRSEC Industry Outreach/Education Activities Survey*, unpublished, November 22, 2004.

through the www.mrsec.org Web site, and have open registration to promote the broadest interactions. Other technical meetings are restricted to participation by companies that are members of an industrial consortium or center at the university. In a number of cases, the MRSEC program is intimately linked with a university center explicitly focused on industrial collaborations. Examples include the Materials for Information Technology Center (MINT) at the University of Alabama, the Cooperative University of Massachusetts Industry Research Program (CUMIRP), the Princeton Institute for Science and Technology of Materials (PRISM), and the Materials Processing Center (MPC) at the Massachusetts Institute of Technology. The explicit linking of a MRSEC with a related industrial consortium program provides good synergy, but it complicates the assessment of the MRSEC industrial collaboration effort. Some collaboration efforts are specifically focused on engaging individual companies. Workshops that are not topically specific but that instead emphasize the breadth of a given MRSEC program have also been conducted as part of industrial collaboration activities. This type of interaction is more typical of a non-thematic MRSEC than of a MRSEC with a strong thematic focus.

Collaborative research projects drive all industrial interactions at MRSECs. As is clear from the annual reports, every MRSEC is able to provide an impressive list of collaborators, including numerous industrial ones. Such industrial involvement provides graduate and undergraduate students and postdoctoral associates in MRSECs with the opportunity to connect their research with industrially interesting problems. There can also be opportunities to work directly with industry R&D staff and/or with managers responsible for product development. Often these collaborations lead to industrial internships. The Ward report provides an analysis of the students performing MRSEC work that involved industrial collaborations.[3] There was a very large number of students working on projects with industry—some of which were entirely supported by MRSEC funding, some entirely supported by industrial funding, and some jointly funded. There are also examples of industry scientists spending time as interns with the MRSEC. This spectrum of interactions provides further evidence of strong engagement between MRSECs and industry.

The committee also saw some very creative approaches to working with industry. A notable example was at the University of Chicago, where MRSEC graduate students have been working with Master's of Business Administration (MBA) students on projects through the Management Laboratory, which is a course run by the Graduate School of Business. By working with students of the School of Business on industrial problems that have both a technical and a business focus,

[3]M. Ward, *MRSEC Industry Outreach/Education Activities Survey*, unpublished, November 22, 2004.

MRSEC students have been offered a unique educational opportunity to expand their appreciation of the role of research in the industrial sector. It is also worth noting that the University of Chicago is an example of a case in which the MRSEC requirement for industrial collaboration created the necessary driver for the center to develop this effort. As noted by its director, Heinrich Jaeger, the University of Chicago did not have a strong history of industrial interactions, unlike many universities (e.g., MIT, the University of Wisconsin, the University of Minnesota). Nevertheless, he stated that the need to have an industrial collaboration effort has been very valuable for students and faculty, especially with respect to informing the research efforts with real problems of interest.

One critical aspect of industrial collaborations is intellectual or proprietary property. The committee discerned from discussing this issue with a number of MRSEC directors that proprietary research with industry is not pursued with MRSEC funding. Such research is directly funded by industry. Some MRSEC directors went to the extent of stating that no proprietary work is done within the MRSEC, since the distinguishing principle of the MRSEC is that all work is shared openly within the center, which is not consistent with conducting proprietary work. A complementary perspective offered by MIT was that if MRSEC work reaches a sufficiently mature point to attract significant external funding, the work is moved out of the MRSEC to make way for new activities. The committee agrees that the philosophy of not doing proprietary work is appropriate and important for MRSEC research.

In any discussion of university-industry collaborations, issues concerning the negotiation of intellectual property rights continue to be a major hurdle for developing stronger and more flexible interactions. While it is outside the scope of this study, the situation with respect to intellectual property rights is in need of serious consideration in order to improve the rate of technical innovation and the transfer of knowledge from universities to industry. MRSECs can largely avoid these concerns by staying away from research that is inherently of a proprietary nature and by working through another entity (e.g., an industrial consortium) to focus on collaborations that are proprietary. As mentioned previously, many universities have very mature industrial consortium programs that readily enable this approach.

The centers are trying to strike a delicate balance between having programs that are compelling for the industrial sector but not so closely coupled with industry that industry is setting the research direction for a center. Having industrial members on the external advisory boards for MRSECs is a common practice to employ for providing a business perspective for the program. Given that an explicit goal for many centers is to have MRSEC research nucleate industry-sponsored research activities and that cost sharing with industrial funding for some centers can be significant, maintaining the research independence of MRSECs is an important goal that requires constant attention. It is important to state that the committee did not find any examples of MRSEC research appearing to be overly focused on the needs of its industrial collaborators.

As industrial collaboration continues to be more important at most research universities, industrial liaison programs have become increasingly coordinated at the university level. Consequently, support is provided by university funds, which are often supplemented by additional funding, such as through state-funded programs. Given the interdisciplinary nature of MRSEC research, the number of faculty involved across campus, and the requirement for industrial collaboration, MRSECs tend to be an important part of the industrial collaboration efforts at their universities. One direct implication of this situation is that the level of MRSEC funding spent on the industrial collaboration effort varies widely from center to center, as shown in the Figure 5.1. In cases where the level of university or state support is sufficient, a significant industrial collaboration effort can be achieved, even with little or no MRSEC funding spent directly on the effort. As an aside, additional analysis showed that there was no correlation between the age of the MRSEC and its industrial collaboration effort's budget, indicating that maturity as a center was not a factor.

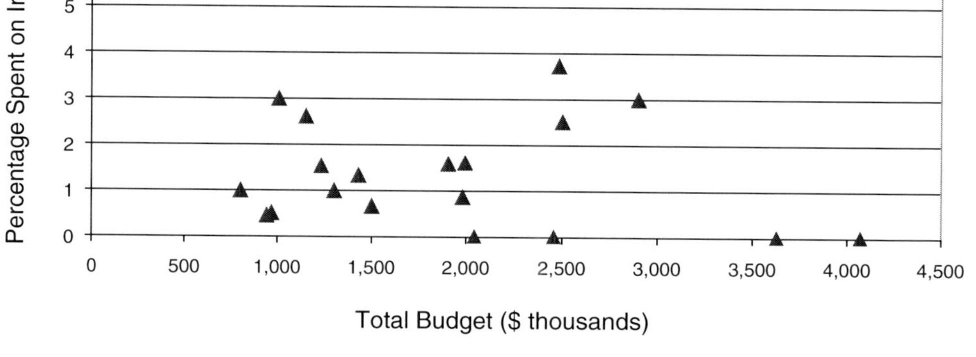

FIGURE 5.1 Materials Research Science and Engineering Center (MRSEC) funds spent directly on industrial collaboration and knowledge-transfer activities as a percentage of total MRSEC budget (by center).

The type of leveraging indicated is typical of how various funding sources are brought together to meet the expectations of different sponsors. The committee is comfortable with this pooling of resources, even if it makes it difficult to understand the specific role of MRSECs in industrial collaboration. An area of concern with respect to financial resources (and time) spent on industrial collaboration activities is associated with smaller centers. Smaller centers, especially if they do not have a strong university-sponsored industrial liaison program, can expend significant resources on this aspect of the program. Consequently, industrial collaboration is one more MRSEC program requirement that can be a proportionately larger burden on small centers as compared with larger ones.

There are a variety of success stories in MRSEC industrial collaboration. Over the years, MRSEC research has led to the establishment of a number of start-up companies. From the committee's evaluation, about 12 start-up companies have been established over a number of years as a direct result of MRSEC work. MRSEC science has also been used by established industries to provide better understanding of their material processes and performance and to help solve problems associated with product development or production. In other cases, access to shared experimental facilities (SEFs) at MRSECs has been seen as critically important to industry, especially smaller companies that did not have certain needed capabilities. It is worth pointing out that as innovation in the United States is being increasingly driven by small (including start-up) companies, an appropriate focus on working with small companies is suitable for the program. The committee also notes that these success stories were largely anecdotal; the narratives generally do not have enough specific information to ascertain the relative importance of the MRSEC contribution.

The committee was generally impressed with the breadth of the industrial collaboration efforts across the MRSEC program. Although some centers showed a stronger focus on industrial interactions than others did, especially based on the type of research conducted, all centers appeared to have a significant effort aimed at meeting the industrial collaboration and knowledge-transfer goals of the program. As industrial liaison programs have become important at research universities, the MRSEC program goals are generally well aligned with the goals of university administration. As previously noted, the explicit MRSEC program requirement for industrial collaboration has been effective at ensuring that all centers give attention to the program's intent with respect to knowledge transfer, even at institutions that do not have a strong history of industrial interactions. The inherent flexibility provided by the NSF program managers in meeting the program goals seems to work well in that centers take different approaches, including some creative ones, to meet the intent of the program effectively.

One potential feature that the committee found notably lacking was interaction among MRSECs in relation to industrial collaboration. There was no evidence of a systematic program or network approach to knowledge transfer, even when programs at various universities could be highly synergetic. A barrier to such interactions is the handling of intellectual property. Nevertheless, if knowledge transfer to spur industrial innovation is a program driver, creating a more effective network among related MRSEC research efforts should be an important goal for the future.

ASSESSING THE EFFECTIVENESS OF INDUSTRIAL COLLABORATION

The committee looked closely at the impact of industrial collaborations with MRSEC partners. This section reviews the committee's data-gathering activities, its analysis, and some perspectives from MRSEC participants.

Methodology

Many direct outcomes of MRSEC industrial collaboration can be used to evaluate its quality and effectiveness. It is clear that a successful MRSEC—that is, one that is reviewed successfully—must address its role in industrial collaboration seriously. Outcomes that can be cited include the following: number of industrial collaborations, time spent by industrial participants on projects, number of MRSEC-funded individuals working with industry, joint publications, patent filings and awards (MRSEC-owned or owned jointly with industry), licensing of patents, and so on. What is not so clear is what criteria (or metrics) should be used to judge the effectiveness of industrial collaboration and knowledge-transfer efforts. Also, it is unclear what metrics are currently being used by NSF to judge the performance of the program as a whole. The adage "What gets measured gets done" is important here and needs to be considered carefully when addressing appropriate metrics for this aspect of the program.

While there are many potential metrics that could be used to assess the effectiveness of the industrial collaboration effort, quantitative information is not typically available on many of the outcomes of potential interest to the committee, and the effort needed to gather it is outside this study's scope. Some quantitative information is available through the MRSEC annual reports, but most of the information collected through all of the sources used by the committee was anecdotal. These sources included teleconferences with MRSEC directors, industrial collaboration coordinators, students, and members of corporate leadership; site visits to MRSECs, Engineering Research Centers (ERCs), and other materials research centers; and formal data requests to NSF and to the MRSECs.

An additional difficulty in developing a clear understanding of the impact from industrial collaboration was that many MRSECs are closely aligned with other complementary centers of research, as mentioned in the previous section. Decoupling outcomes from MRSEC activities with those supported from other sources is impossible, and determining the value of attribution to a particular outcome (such as a patent) could be misleading.

In considering criteria and metrics for assessing effectiveness, there was a desire to develop a systematic approach for assessing the impact of specific centers, which speaks to review criteria, and how to assess the impact of the MRSEC program as a whole.

Analysis of Data

The MRSEC annual reports do contain some quantitative information on accomplishments that can be indicators of the impact of industrial outreach. This section provides a brief summary and analysis of the MRSEC program's performance based on data reported in annual reports (generally from 2005) and other sources noted.[4]

In recent years, many universities have increased their patenting activity. Pressure has increased at these institutions to convert their research into potentially profitable intellectual property (IP). Although most patents generate no direct income, there are examples of license royalties that are valuable to universities. Given the charter for industrial collaboration and the interdisciplinary nature of MRSEC research, it is worth examining whether MRSECs are more successful at generating IP than is generally seen within the academic community.

The program's contribution to patents awarded to academia has hovered around 1 percent since 1999 (Figure 5.2), where academia's share of total U.S. patenting activity is about 4.4 percent over the past 5 years. At an average of 0.21 percent of federally supported university R&D expenditure, the MRSEC program secures more IP per dollar spent than the average university R&D dollar. However, when examining patent filings and patent awards within the program from center to center, the committee saw no correlation between the level of industrial collaboration (measured by number of collaborations, funding, and so on) and the IP activity. Several MRSECs had a significant level of patent activity, but most centers had little or no patent output. It may be that a center's patenting activity is more firmly rooted in the university's culture and emphasis on IP and licensing than being related to its success in industrial collaboration and knowledge transfer. Differences in internal university policy and state law exist also across MRSECs, which can affect patenting activity. In addition, it is not clear whether university-held patents have a beneficial impact on industry and the level of knowledge transfer.

[4]See, for example, Harvard University's Materials Research Science and Engineering Center, *2005 Annual Report*, unpublished, 2005.

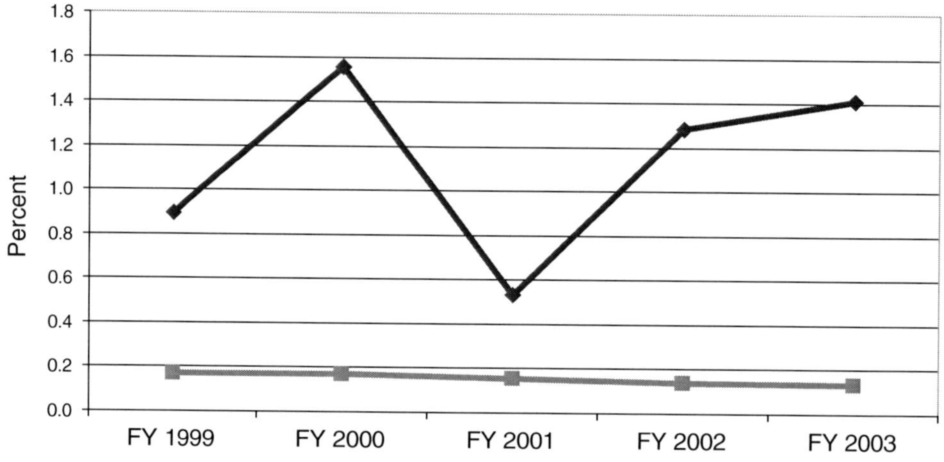

FIGURE 5.2 Percentage of patents (top curve) and academic research and development funding (bottom curve) belonging to Materials Research Science and Engineering Centers (MRSECs) as a percentage of the total for universities, FY 1999 to FY 2003. SOURCE: National Science Foundation, *Science and Engineering Indicators 2006*, and Division of Materials Research, National Science Foundation.

Consequently, patent output, while reported annually, was not seen as an especially useful metric for determining programmatic success.

Another potential metric of the effectiveness of industrial outreach activities at a MRSEC is the number of industrial collaborators involved. The number of industrial collaborations for a MRSEC is reported annually. However, it is difficult to assess the significance of the collaboration on the basis of the information available. The time or resources applied to a collaboration may provide some insight into the quality of the interaction, but there is no straightforward way to assess the quality of a collaboration, especially as it relates to impact. There is also no obvious correlation between the level of MRSEC funding for industrial outreach and the number of industrial collaborations.

Successful interactions might also be expected to provide return on investment through company-sponsored research (see the subsection below entitled "Industrial Perspectives" for an extended discussion of the industrial view of MRSECs). Some MRSEC programs are very clear about their goal of obtaining complementary research support from industry; others are not as focused on this goal. One difficulty in assessing impact under this category is that support provided by industrial sponsors is generally brought in through direct contracts with individual faculty or through one of the complementary industrial liaison centers on campus. Consequently, accomplishments in this category may not be attributed to the role of the MRSEC.

Additionally, joint industry–university publications could indicate that a MRSEC has a healthy industrial collaboration program. However, industry does not value publications like university departments do, and the mismatch of motivations becomes a problem with considering this outcome as a metric for success.

Furthermore, successful university research initiatives, depending on their character and the research topic, can develop into small "spin-off" companies. However, these spin-offs occur somewhat infrequently and depend heavily on local circumstances. While success in creating spin-offs is a positive outcome, this metric is probably not an especially important one for all centers.

The use of SEFs at a MRSEC is also another metric of impact. The use of facilities tends to be by local companies, often smaller ones, that cannot or do not want to invest in highly specialized equipment. The use of facilities, especially when coupled to a collaborative effort, is a positive outcome. It was noted by MRSEC directors that the use of MRSEC facilities is appropriate under many circumstances, but not for doing a lot of routine work for a company. Routine access to university facilities is better accommodated through an industrial liaison program.

Intellectual property, collaborations, joint publications, creation of spin-off companies, direct funding, and other quantitative outcomes may collectively provide an indication of the health and vitality of a MRSEC's industrial collaboration efforts. All of these metrics are helpful in understanding industrial impact, but it is important not to rely only on simple metric(s) to judge effectiveness lest the overall picture of program success be lost.

MRSEC Perspectives

MRSECs conduct a very wide variety of research, differing not only by topic but also by degree of "applicability." Some centers' research is more thematic, or focused on a particular problem, whereas others have multiple thrusts without strong connection among the Interdisciplinary Research Groups (IRGs). MRSECs are also judged on the basis of being centers of excellence in basic research with a charter to focus on furthering the state of knowledge in materials science, rather than focusing on the needs of industry, which is appropriate. Nevertheless, an active industrial partnership effort has a positive impact on the research in that industrial challenges often stimulate new science. Consequently, knowledge transfer goes in both directions, and MRSECs see the benefits of this exchange. Some MRSECs even credited the program requirement for collaboration as being the impetus for their establishing a valuable mechanism for knowledge transfer.

Although many centers see industrial collaboration and technology transfer as generally beneficial, and NSF stresses it in the proposals, according to many

MRSECs, industrial outreach receives an incommensurately low amount of attention in the review process, or in some cases, is ignored. There may be several reasons for this situation or impression. First, some MRSECs find that the NSF-selected review panels are not populated to evaluate these activities astutely and do not regularly include members from industry (the committee understands that this situation has been improving with recent reviews). Without an appropriate industrial perspective, it can be difficult to judge whether an industrial partnership program has fulfilled its goals. Second, the review panels do not know how to evaluate and assess industrial collaboration because the NSF does not provide them with a set of criteria to use. Without clear criteria, it is hard for review panels to objectively evaluate programs about which they know little.

Students are often cited as the most important aspect of a MRSEC. Beyond providing financial support for students and postdoctoral associates, MRSEC participation is believed to provide a unique and broader research experience than would be possible under a single-investigator grant. MRSECs provide more industrial interactions for students on some campuses than would otherwise be the case, and MRSEC involvement can often lead to other opportunities, such as industrial internships. Since MRSECs put a concerted effort into stimulating industrial partnerships, one might expect this emphasis to have an impact on students' employment decisions. Interestingly, as shown in Figure 5.3, a student receiving a Ph.D. from a MRSEC is equally as likely to take a job in industry as any other student with a degree in materials science and engineering nationwide. It is tempting to conclude that the industrial collaboration part of the MRSEC experience has little impact on the career decisions of those trained in materials science.

Industrial Perspectives

The committee found that the view of MRSECs from the industrial perspective was quite mixed. Since the main goal of the MRSEC program is to carry out fundamental materials science that is not directly tied to industrial interests, perhaps it is not surprising that the perceived interest in the MRSECs by industry is modest. However, it is important to keep in mind that Figure 2.3 showed that industrial support of all academic research and development is quite modest. While the MRSECs can list an array of successful interactions with industry, their direct impact on the development and application of new technologies would appear to be quite limited.

Smaller companies can clearly benefit in straightforward ways from MRSECs—for example, through access to equipment and capabilities that such companies could not afford to purchase themselves. Moreover, access to expertise in particular areas of materials research is facilitated by the MRSEC through a single focal contact point that provides access to a larger number of academic re-

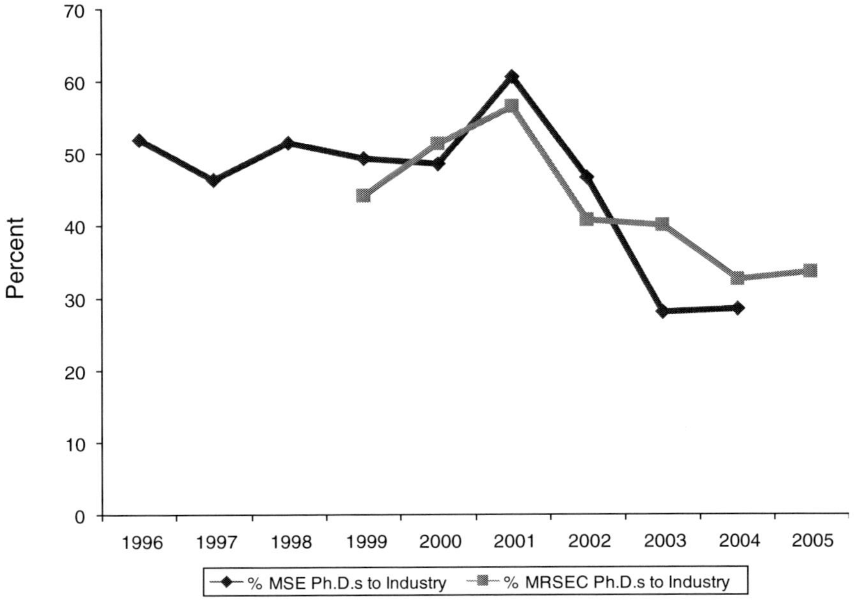

FIGURE 5.3 Percentage of new Ph.D.s with materials science and engineering (MSE) degrees who are then employed in industry compared to the percentage of Ph.D.s awarded their degree by a Materials Research Science and Engineering Center (MRSEC) who are then employed in industry, 1996-2005. SOURCE: Division of Materials Research, National Science Foundation.

searchers. While smaller companies may have more to gain from the MRSECs than larger companies might, it is clearly more challenging for the MRSECs to identify and develop interactions with a multitude of small enterprises. On the other hand, larger companies have less direct interest in the MRSECs on a day-to-day basis, but MRSECs can identify these larger enterprises more readily and perhaps see these companies as being a more likely source of additional funding.

One potentially important role of the MRSECs is the training of students in the methods of working in large interdisciplinary research projects, perhaps more similar to the style of industrial research. However, as discussed above, there is no evidence that students who carry out their Ph.D. research in a MRSEC are more likely to find a job in industry.

In the past two decades, the large industrial research laboratories in the United States, which used to carry out significant broad-based exploratory research programs in materials, have largely disappeared as increased globalization and worldwide competition makes them uncompetitive. Thus, programs such as the MRSEC program become of greater importance for the long-term sustainability of technology leadership in the United States. This need is recognized by industry, which largely supports the scientific independence of the MRSECs. Paradoxically, too great an influence by industry would rob the MRSECs of their very importance to industry. However, strong links that foster the transfer of knowledge from the MRSECs to industry are of paramount importance, but these links appear to be relatively weak. For example, the committee found very few examples of scientists from industry spending any significant time at a MRSEC. By contrast, this is common practice in Japan where scientists from the major electronics companies such as Hitachi, Toshiba, NEC, SONY, and other companies post their scientists to major universities and national laboratories for one or more years at a time. This practice is also common in Europe where Hitachi, for example, has several small exploratory research laboratories embedded on university campuses in several countries. Encouraging extended sabbaticals by industrial scientists to spend time within a MRSEC might be a productive means of enhancing links between MRSECs and industry.

Major companies are willing to spend significant sums of money (relative to an individual MRSEC funding) to encourage exploratory research in broad areas of interest to the company, but because the landscape of industrial research has changed, they now need to look outside their walls to support it. One industrial partner told the committee that it was much cheaper for his company to fund basic research at a university than to carry out the same research in his own research laboratory. This gives the company flexibility to approach more short-term, applied research problems while leaving the longer-term, broader-scope research to the university-based MRSEC. Given this relationship and the static funding of the MRSEC program over the past several years, an initiative to attract increased funding from industry to support the MRSEC program would appear to be an attractive proposition. Such an initiative would likely be most successful if carried out by an industry coordinator at NSF for the network of MRSECs as a whole.

Only one MRSEC, the Center on Polymer Interfaces and Macromolecular Assemblies (CPIMA) program centered at Stanford University (with partners at the University of California at Davis and at Berkeley), has a full industrial partner, the IBM Almaden Research Center. CPIMA appears to have an outstanding record of accomplishment with notable scientific successes. The committee sees that such partnerships can be very valuable and recommends that such partnerships be encouraged.

NSF Perspective

The NSF sees the MRSEC program performing exceptionally in enhancing industrial outreach and knowledge transfer. In discussions with officials at NSF, it was evident that there was a clear understanding of the tangible benefits produced by the program. In response to the committee's query, "What are your program goals for industrial interactions?," NSF MRSEC program managers gave the following perspectives:

- Dissemination of knowledge, more multifaceted than individual principal-investigator efforts;
- Enhancing the educational experience—educating students by providing opportunities for industrial collaborations;
- Leveraging NSF funding of centers through industry support and projects;
- Intellectual property—patents and licensing;
- Contributing to the establishment and success of new businesses;
- Being a national resource, especially by making unique facilities available; and
- Informing the research—using industrial challenges to help catalyze the formulation of research directions within the MRSEC.

This perspective is based on a belief that industrial interactions allow a center to magnify its impact in a way that is greater than the impact that an individual could create. This broader impact can occur for several reasons:

- As centers, MRSECs act as nucleation points for contact with industry with a scale and focus that matches well with business-driven research;
- Industry interactions enhance the experience of the students involved;
- Collaborating with industry can attract new and other types of resources to the center; and
- Industrial partners benefit from economic and competitive advantages.

MRSECs are engaged in collaboration to varying degrees, numbers exist to show the activity, and centers report a positive effect. What is missing is a critical understanding of impact, partly because of the difficulty with measuring impact as a desired outcome. Resolving this dilemma will require a concerted effort by NSF, since it is difficult to find reliable, standard metrics to evaluate impact with a program as diverse as MRSEC.

In addition, the requirement to have an industrial collaboration effort is just one of many program expectations that must be satisfied simultaneously. It was also not evident that the NSF understood the impact of numerous program requirements, including this one, on the centers, especially the smaller ones. This issue associated with balancing multiple program goals without clear relative priorities may explain why NSF believes that while industrial partnership is an important part of the program, a struggling industrial outreach program will not sink a MRSEC. Nevertheless, NSF needs to determine what it expects out of the MRSEC program, determine how to assess those expectations, and convey that information explicitly to the review panels to improve the process.

The MRSEC program is one of many mechanisms that a research center can use to conduct industrial collaboration and knowledge transfer. Other center programs at NSF, such as the Science and Technology Centers (STC) and the Engineering Research Centers (ERCs), also have requirements for industrial outreach.[5] Science and Technology Centers are problem-driven NSF research centers investigating topics ranging from remote sensing of ice sheets to environmentally friendly solvents. An STC is composed of a multi-university collaboration of researchers investigating a single problem. STCs are required to "have significant intellectual exchange and resource linkages among various types of institutions and organizations to facilitate knowledge transfer," and to "include industrial, national or international internships or other career broadening experiences as appropriate to the research are."[6] ERCs are tightly linked to industry and attack research problems with a specific end application in mind. ERCs are expected to create an "interdisciplinary research environment where academe and industry join in partnership to advance fundamental knowledge and engineered systems" and "are expected to be self-sustaining after ten years when NSF support ceases" by industry support and other means.[7]

There are important differences between these centers and MRSECs with respect to industrial collaboration. MRSECs are research-driven, and industrial outreach should be a natural outcome of the research focus. The other centers are tightly linked to application outcomes, and industrial collaborations are integral to success. In this context, maintaining the MRSECs' focus on basic research is viewed by the committee as highly desirable.

FINDINGS AND RECOMMENDATIONS

Conclusion: The program goals for MRSEC industrial collaborations are appropriate. A flexible approach to meeting those goals is essential to address the needs and capabilities of the individual MRSECs.

Conclusion: The MRSEC program requirement for industrial collaboration leads to important activities that likely would not occur otherwise

[5]See http://www.erc-assoc.org/ and http://www.nsf.gov/od/oia/programs/stc/ for additional information.

[6]National Science Foundation, *Program Solicitation for Science and Technology Centers: An Integrative Partnership*, NSF 03-550, Washington, D.C., 2003.

[7]National Science Foundation, *Program Solicitation for Engineering Research Centers: Partnerships in Transforming Research, Education and Technology*, NSF 07-521, Washington, D.C., 2007.

(e.g., workshops, short courses, external advisory boards with industrial advisers).

The MRSEC directors whom the committee informally interviewed all were supportive of the industrial outreach and knowledge-transfer goals for the program. Although some centers had an existing campus culture that already supported industrial outreach activities, other MRSECs had to create a culture of industrial outreach in order to respond to program requirements. As a result, all centers had substantial collaboration efforts that added significant value to the overall program. The committee found that local flexibility in meeting the program goals was effective in taking advantage of inherent differences between MRSECs, the university environment that they resided in, and the targeted industrial community. As with education and outreach, there is a disproportionate impact on small centers to demonstrate accomplishments in all MRSEC program goals.

Conclusion: MRSECs have developed industrially relevant programs while maintaining a commitment to solving long-term research problems.

Maintaining this approach is important to the quality of the research efforts and to educational continuity for students, especially those involved in Ph.D. research programs. Industrial interactions are a positive part of the educational experience for students. The ability to connect their research to external needs and to have an opportunity to work with industrial scientists was clearly cited as being beneficial by the students interviewed by the committee.

MRSEC research programs are stimulated through industrial interactions as a result of the challenges and research needs articulated by industrial partners. This positive feedback to the research planning effort was reinforced through discussions with numerous MRSEC directors. To date, MRSEC industrial collaboration appears to have been primarily focused on large industrial research laboratories, but the opportunity to interact more with innovative small and start-up companies is coming to be appreciated to a greater extent.

MRSEC research is generally well focused on leading-edge and transformational research, as appropriate. MRSECs have developed industrially relevant programs without getting involved with near-term problem solving.

Conclusion: MRSEC industrial collaboration efforts are generally supported by multiple sources, in addition to MRSEC funds, such as funds from industrial partners themselves.

In a few cases, a significant portion of the MRSEC funding (more than 8 percent) was used for industrial outreach. More typically, MRSEC industrial partnerships are supported primarily by university and/or state funding and are usually assisted by a university liaison program. This leveraging is valuable to the MRSEC program in meeting its goals, but it makes assessing the effectiveness of the industrial outreach program more difficult to judge as a function of MRSEC resources supporting the effort.

Conclusion: The importance given industrial collaboration and technology transfer in the review process is seen as not being commensurate with the importance of this program goal.

Industrial outreach is seen as not having an emphasis in evaluating MRSEC performance that is commensurate with the importance of this program goal. The evidence on this point is all anecdotal (from conversations with MRSEC directors). Nevertheless, the strong impression is that a viable industrial collaboration effort is required for a successful renewal, but an especially strong outreach effort is not rewarded. This impression is consistent with minimal industrial involvement on MRSEC review panels. One aspect of this finding is that the NSF struggles with being able to assess the effectiveness of the industrial outreach effort. If the program managers cannot clearly articulate expectations and how to evaluate performance against those expectations, it will be almost impossible to improve the way in which this aspect of MRSEC performance is considered as a part of the reviews.

Each MRSEC tends to have its own program for industrial outreach and collaboration, and industrial contacts typically do not interact with more than one MRSEC. There is evidence of occasional industrial interactions that incorporate more than one MRSEC, but collaborative efforts among centers are the exception.

MRSEC leaders understand the change in the research landscape within the United States and are trying to respond appropriately. In particular, there is a shift away from a system dominated by several large, comprehensive industrial research laboratories toward a greater number of small and entrepreneurial companies involved with technology innovation. Understanding how to work effectively with these smaller companies and ensuring that these interactions are properly recognized and valued by the MRSEC program will be critical.

The committee was generally impressed with the breadth of the industrial outreach efforts across the MRSEC program. Each center seems to have a vital industrial outreach activity that meets the stated program goals. While it is difficult to evaluate the impact of the industrial outreach efforts clearly, the committee believes that the MRSEC program is generally meeting its goals and that the industrial outreach is valuable.

Recommendation: NSF should establish metrics for evaluating the effectiveness of industrial collaboration and technology transfer.

In addition to considering worldwide best practices, NSF should quantify the relative importance of industrial outreach and knowledge transfer relative to other program requirements in program solicitations. This would enable centers to put the appropriate focus and resources on this aspect of their center and for reviewers to make appropriate judgments about accomplishments.

> **Recommendation: Together with the team of MRSEC directors, NSF should provide a mechanism to enable industry to effectively understand the resources and expertise available through the network of MRSECs. This may require a coordination function that currently does not seem to exist, such as a national network liaison officer based at NSF.**

Industrial outreach and knowledge-transfer effort is inherently based on interactions among people. Encouraging more personnel exchanges, such as student internships, extended sabbaticals for industrial researchers at MRSECs, visits by MRSEC faculty to key industry partners, significant industrial involvement on MRSEC advisory boards, and so on, will be essential to effective knowledge transfer and skill development (especially for students). For instance, centers that have better exposure to industrial partners could provide access to students involved in other MRSECs. Centers' tapping into these shared opportunities presented by the entire MRSEC program would enhance the program's overall impact.

6

The Future of National Science Foundation Materials Centers

After its examination of the history of the MRSEC program and its impacts, the committee formed several judgments about the future direction of the program.

PERCEIVED AND MEASURED IMPACT OF MRSECs

Why do outstanding people and institutions pursue Materials Research Science and Engineering Center (MRSEC) grants with all of the associated responsibilities? An analysis of inquiries made of faculty at both MRSEC and non-MRSEC institutions revealed multiple motivations for participation in the MRSEC program.

Conclusion: MRSEC awards continue to be in great demand. The intense competition for them within the community indicates a strong perceived value. These motivations include:

- **The ability to pursue interdisciplinary, collaborative research;**
- **The resources to provide an interdisciplinary training experience for the future scientific and technical workforce from undergraduate to postdoctoral researchers;**
- **Block funding at levels that enable more rapid response to new ideas, and that support higher-risk projects, than is possible with single-investigator grants;**
- **The leverage and motivation MRSECs provide in producing increased institutional, local, and/or state support for materials research;**

- **The perceived distinction that the presence of a MRSEC gives to the materials research enterprise of an institution, thus attracting more quality students and junior faculty; and**
- **The infrastructure that MRSECs can provide to organize and manage facilities and educational and industrial outreach.**

These factors suggest that there are strong positive influences of the MRSEC program on the conception of research ideas and the ability to pursue them quickly and effectively, which in turn have clear, positive implications for maintaining and advancing U.S. research competitiveness in the materials field. This observation must be tempered in the context of the current funding situation, in which MRSECs are asked to take on increasing responsibilities without the availability of commensurate resources.

Conclusion: The committee examined the performance and impact of MRSEC activities over the past decade in the areas of research, facilities, education and outreach, and industrial collaboration and technology transfer. The MRSEC program has had important impacts of the same high standard of quality as those of other multi-investigator or individual-investigator programs. Although the committee was largely unable to attribute observed impacts uniquely to the MRSEC program, MRSECs generally mobilize efforts that would not have occurred otherwise.

MRSECs conduct and publish research with performance characteristics similar to those of other programs. The committee came to believe that MRSECs enable the formulation of some research activities that would not have occurred outside the program. The shared-acilities element of MRSECs has very high value because it represents a significant portion of the National Science Foundation (NSF) investment in midsize facilities for materials research. The MRSEC education and outreach programs clearly benefit from the sharing and pooling of resources; improvements by NSF and the participating communities are needed, however. Although the industrial collaborations that take place within the MRSEC framework are of a character similar to those conducted elsewhere, the activities initiated by MRSECs generally represent efforts that would not have occurred otherwise.

The MRSEC program allows NSF, and thereby the nation, to make a different style of investment in materials research: one that couples group-based research with facilities, industrial interactions, educational programs, and so on. Thus, from the standpoint of diversity of funding mechanisms, if the MRSEC mechanism produces equally high quality results, retaining it enhances the resilience of the overall portfolio.

The committee formed several other impressions quite strongly as a result of its site visits, testimony at meetings, and in its discussions. The committee was unable to construct a method for developing quantitative evidence to substantiate these expert judgments, however.

- Interdisciplinary, group-based research that includes access to facilities that cannot be supported by individual investigators is critical to the progress of materials research.[1]
- "Local management" permits a more flexible and responsive approach to the local environment. Although the MRSEC award is identical in structure at the highest levels across all institutions, the specifics vary widely from center to center. The committee observed that this was primarily because of differences among the campus cultures, university administrations, faculty personalities, and to some extent, state and other local oversight and funding bodies. By delegating authority to each center, MRSECs are better able to take advantage of their local circumstances, including negotiating with the campus or state authorities for in-kind contributions.
- MRSECs are an opportunity for flexibility not possible in other funding mechanisms. The 6-year funding cycle and seed program promotes basic research that may not show immediate payoff, and high-risk/high-reward research; however, MRSECs appear to be moving toward greater uniformity (in size and topics), and change is usually found only during recompetition.
- The long-term nature of MRSEC support is a great advantage.[2] In addition to funding basic science, which is essential to the progress of the field, graduate students have a 5-year lifetime. The vagaries of support can impede their progress. If science alone were driving the evolution of Interdisciplinary Research Group (IRG) topics, one would expect to have a continuous rate of IRG turnover MRSEC-wide each year.
- The lack of mechanisms to support the purchase and maintenance of research equipment and the training of students on this equipment is troubling. Instrumentation programs generally support equipment, but not the infrastructure necessary to hold it together. The National Research Council (NRC) report *Midsize Facilities: The Infrastructure for Materials Research* stresses this point, stating, "A continuing and fundamental challenge facing a majority of small to midsize facilities is planning, securing, and maintaining the long-term infrastructure necessary for productivity

[1]See, for instance, National Research Council, *Facilitating Interdisciplinary Research*, Washington, D.C.: The National Academies Press, 2006.

[2]National Research Council, *Midsize Facilities: The Infrastructure for Materials Research*, Washington, D.C.: The National Academies Press, 2006, p. 188.

> and success."[3] As documented in that report's appendixes, programs such as the NSF-wide Major Research Instrumentation (MRI) program and the NSF's Division of Materials Research (DMR)-specific Instrumentation for Materials Research (IMR) focus on providing assistance for the acquisition of instrumentation.

The committee came to unanimous agreement that a critical strength of the MRSEC program is the relatively autonomous management of each center, so-called local management. The committee believes that by encouraging each center director to steer his or her center toward topics and resources that make optimal use of the local institutional environment, NSF has significantly enhanced the MRSEC program. That is, by encouraging and supporting "local management," MRSECs have avoided some key pitfalls of the "one size fits all" management rubric.

The MRSEC program is unique in its lack of a formal sunset clause; although centers lack certainty beyond the horizon of their current award, they may compete for renewal an unlimited number of times. Although the committee could not document the impact of this policy on the research results of MRSECs, it became convinced that the policy added an important dimension to the overall portfolio of DMR investments. For instance, the fact that some MRSECs are sited at institutions with involvement dating back to the Materials Research Laboratories (MRLs) and the Interdisciplinary Laboratories (IDLs) is not a sign of entitlement. The people, ideas, and tools of 1960 would never win a present-day competition for a MRSEC. These types of legacies are really testimony to the ability to reinvent one's self and remain competitive.

CHALLENGES GOING FORWARD

The previous chapters of the report have established the overall value of the MRSEC program; however, they also have raised the critical problem that the evolution of the MRSEC program, both in numbers of centers and in the set of required responsibilities, has not been matched by commensurate funding. The number of MRSECS has expanded from 10 to 26, and the 26 MRSECs of today have a much broader and more diverse mission and scope that mandate educational and industrial outreach. In addition to lack of growth in as-spent funding, essentially every class of direct and indirect research cost has grown. Funding levels have failed to keep pace with this inflating cost basis—whether in the context of student or postdoctoral stipends, tuition rates, or the cost of capital equipment and supplies. The increased number of MRSECs being supported only amplifies the strains.

[3]National Research Council, *Midsize Facilities: The Infrastructure for Materials Research*, Washington, D.C.: The National Academies Press, 2006, p. 188.

Current MRSECs are smaller in actual and constant dollar terms and are expected to accomplish more than the MRL programs that they replaced.

Conclusion: The effectiveness of MRSECs has been reduced in recent years as a result of increasing requirements without a commensurate increase in resources. Increasing the mean grant size is necessary to allow the program to fulfill its important mission goals.

In addition to increasing industrial collaboration and education and outreach (EO) responsibilities,[4] the number of MRSECs has increased while the MRSEC program has remained at a relatively constant budget level. Average funding for centers, in constant dollars, has decreased substantially in the past decade. Declining funding has been particularly detrimental to building and maintaining the advanced instrumentation necessary for leading-edge materials research. Another decade of similar decreases will undermine the future contributions of the MRSEC program.

In flat funding environments, and whether explicitly mandated or not, these embedded cost pressures can only be met through reductions in the scope of the programs. These reductions have come in many forms. First, MRSECs are losing their capacity to develop, manage, and, most importantly, to sustain state-of-the-art experimental and computational facilities for materials research. The loss of this infrastructure will have damaging consequences on the competitive standing of the United States in this critical area of physical science—one that underpins technologies that are critical to both prosperity and national security. The decreasing purchasing power of MRSEC funding—in the context of supported research personnel—must also have collateral impacts on staffing levels. Paradoxically, self-reporting suggests that the numbers of students and postdoctoral fellows supported by MRSEC funding are in fact increasing—doing so even in institutions that are flat-funded over the 6-year term of the grant. This highlights the crucial role that so-called funding synergies (leveraging) have come to play in this program. The MRSEC program neither fully funds nor does it wholly own the creative outputs of its various programmatic components, and yet it continues to justify its existence by contending that it is in some way different from the individual principal-investigator (PI) grants with which it competes for funds.

Strains on the MRSEC program have potentially serious consequences. When resources are scarce, risk taking and innovation are the first to suffer. MRSECs

[4]An examination of the MRSEC program solicitations and reporting guidelines reveals a number of escalating requirements placed on successful centers. Requirements for activities to recruit and promote workforce diversity, as well as junior faculty development, expanded in 2007; international activities were strongly encouraged in 2004. More important, however, is the impact of increasingly fierce competition for the MRSEC awards. To remain competitive, proposals must promise to do more and more with resources that are steadily eroding through inflation.

offer a pooling of resources to enable some hedging of bets on creative research with more standard investigations. But when human and financial resources are in short supply (or even fail to grow with inflation during a 6-year award cycle), the less-certain research is scaled back. While the committee did not find persuasive evidence that MRSECs have reached this breaking point, it was clear that some of the smaller centers are struggling.

Perhaps the greatest challenge going forward for the MRSEC program is the relative inability to quantify its unique value. The 1976 MITRE Corporation panel failed to ascertain the unique characteristics of the research results enabled by the MRLs (see the section entitled "MITRE Report" in Chapter 2). This present committee has not been able to identify a set of performance indicators for the MRSEC program. This does not suggest, however, that the program is without value. On the contrary, the increasing competition for MRSEC awards suggests that the value is quite high and that it is simply too complex to measure with just a few parameters. This state of affairs is not unusual in science, however: the peer-review system is the most commonly used assessment tool for evaluating past performance and projecting future results.

The MRSEC program is thus at a critical point in its history. The current trends suggest that, if left unchanged, the capacities and competencies of the centers will be subject to both relative and absolute decline. Centers will have to be still smaller, operating programs of research that have a lesser reach than the programs of the original Materials Research Laboratory system that they replaced. To the extent that facilities can be supported, they will likely fail to rise either to state-of-the-art levels or the standards being set by global competitors. These trends, if left unchanged, suggest a program that will not be able to make significant contributions to the national portfolio of materials research: a program that does many things, but excels at none of them.

A NEW LOOK

Although many positive outcomes have been identified in this report, it is the committee's judgment that the resources are simply too small and are spread over too many centers to enable the MRSEC program to continue to have substantial impact in research, facilities management, and education and industrial outreach. The downside of local management is that NSF has not specified clear, overarching objectives for the program or any of its components (education, industrial outreach, and so on). The overall coherence of the program suffers as a result.

MRLs, and later MRSECs, were conceived to be centers where interdisciplinary groups of materials researchers were brought together around enabling infrastructure, including the student and postdoctoral support that would allow them to tackle long-term, significant problems. Despite the substantial leveraging of institutional support, only the largest MRSECs have sufficient funds to purchase, maintain, and staff significant shared experimental facilities (SEFs). Student and postdoctoral support levels are now well below one student per investigator, and faculty are discouraged from taking summer salary. Investigators must increasingly combine resources to conduct research at MRSECs, making it difficult to identify "MRSEC research."

The MRSEC program can and must play an important role in the nation's material research efforts; however, more effective leveraging of funds is necessary. Incremental change will not be sufficient, and the committee proposes here a restructuring of the MRSEC program that will both preserve its original character and provide greater research flexibility.

Conclusion: The MRSEC program needs to evolve in order to successfully meet its objectives in the coming decade. To do so, the National Science Foundation must restructure the program to reduce requirements, reduce the number of MRSEC awards, and/or increase the total funding of the MRSEC program while preserving its positive elements.

Given the multiple demands on MRSECs, the program has been underfunded for some time, and the situation has been getting worse. One solution is to increase the level of funding of the MRSEC program, perhaps justified on the basis of proposal pressure and the importance of the field to the nation's economic and strategic security. Whether or not new resources become available, the committee recommends a mix of large, well-funded centers and small, appropriately funded research groups.

Recommendation: To respond to changes in the budgetary landscape and changes in the nature of materials research in the coming decade, NSF should restructure the MRSEC program to allow more efficient use and leveraging of resources. The new program should fully invest in centers of excellence as well as in stand-alone teams of researchers.

Resources for basic research, especially materials research, have not kept pace with overall economic growth in the past decade. Expectations regarding the range and extent of impacts enabled by NSF's programs have also changed. And materials research has continued to mature as a discipline. The MRSEC program can be positioned to better facilitate its contributions in the next decade by improving the focus of its resources on targeted, specific objectives and by increasing its flexibility to allow specialization based on individual strengths. The committee developed one detailed vision for achieving these objectives; it is articulated here.

The current MRSEC program could be evolved beyond the current model, by which "each center should try to do everything." Units of the program would be encouraged to focus either on agile teams of group-based research or on larger centers of excellence that pair research teams with additional resources for facilities and outreach. In practice, one approach to this is splitting the MRSEC program into two parts: one part of the available funding should be invested in a small number of larger MRSECs called Materials Centers of Excellence (MCEs), with ample infrastructure, and the other part of the available funding should be invested in establishing smaller-scale Materials Research Groups (MRGs). The committee does not want to be too prescriptive, but a first guess might be rough equal parts. A general decline in resources, coupled with increasing requirements, has made it impossible for MRSECs to meet all of the program expectations well; therefore, it makes sense to fund fewer centers at a larger dollar amount per center. MRGs would fund high-quality research requiring less infrastructure and satisfying the usual NSF review criterion for broader impacts. The committee believes that this recommendation is valid even in the favorable event of an overall increase in MRSEC funding.

This prototype program structure is described in Tables 6.1, 6.2, and 6.3 using the assumption of $30 million per year for the MCE program (perhaps 10

TABLE 6.1 Comparison of the Current MRSEC with the Possible New Materials Center of Excellence and Materials Research Group Programs

	Annual Budget		
Category	Existing MRSEC	New "MCE"	New "MRG"
Budget	$1 million-$4 million	$3 million-$5 million/yr	$0.5 million-$1.0 million
Equipment	$0-$1 million	$1 million + yearly operating costs	$0.1 million-$0.2 million equipment
Review cycle	5 year + 1 year	5 year + 1 year	4 year + 1 year
Number of awards	26 total	10 to 12 total	45 to 50 total
Number of awards/cycle	11 renew/2 (last cycle)		15/competition
Proposal evaluation	Preliminary proposal and reverse site visit	Preliminary proposal and reverse site visit	Panel selection (no reverse site visits)
Theme	Central theme	Unifying theme or facilities	Single theme
Multi-institutional	Mostly single campus	Maybe multi-campus	Many would be multi-campus to take advantage of expertise
Educational outreach (EO), industrial collaboration and outreach (IO)	EO/IO required	REU required, IO required; can compete for independent EO and/or IO supplements of up to $1 million	None required
Management	Director	Director	PI

NOTE: MRSEC, Materials Science and Engineering Center; MCE, Materials Center of Excellence; MRG, Materials Research Group; REU, Research Experiences for Undergraduates; PI, principal investigator.

TABLE 6.2 Example of an Annual Budget for a Materials Center of Excellence

Category	Annual Budget (U.S. $)
Facilities • Equipment: $1,000,000 • Staff: $1,000,000 • Maintenance: $200,000	2,200,000
Students and postdoctoral associates (one per investigator)	1,500,000
Outreach (education and industry)	200,000
Seed/discretionary	500,000
Intercenter brainstorming	25,000
Administrative support	100,000

TABLE 6.3 Example of an Annual Budget for a Materials Research Group with Five or Six Investigators

Category	Annual Budget (U.S. $)
Students and postdoctoral associates	500,000
Seed/discretionary	100,000

such centers) and $30 million per year for MRGs (perhaps 50 such groups). The $30 million figures were determined assuming that the MRSEC program needs a minimum of 10 or 11 centers for critical mass. This proposal, therefore, is initially revenue-neutral, but it would support scaling to higher levels of investment.

The committee considered drafting a full Request for Proposals as an exercise, but, recognizing the expertise and wisdom of NSF program directors and their formal advisory committees, it chose not to prescribe an explicit framework and so only provides here an illustrative outline of the appropriate level of effort and texture of the two new program elements.

Materials Centers of Excellence

There is ample evidence that national investment in infrastructure for materials research has been woefully weak for the past generation, and not only in the context of the MRSEC program. Under this proposed new program, approximately half of the budget would go to Materials Centers of Excellence, with program-wide infrastructure concentrated in these centers. The MCEs would have much the same mix of activities as expected for a current MRSEC: excellent and focused research, a compelling interdisciplinary environment for student training, powerful research tools, sustained educational outreach, and responsiveness to industrial needs. By suggesting a concentration of more resources in these MCEs, the committee's intention is to ensure an appropriate level of funding for this broad and diverse mission. These MCEs would be similar to existing MRSECs, with the following important differences: The MCEs would have to provide accessibility and support for researchers from their own institution and from other institutions. MCEs might evolve into regional centers, thus expanding the materials research infrastructure for many researchers, not just those involved in the MCE. While the MCEs would consequently serve partially as user facilities, much as the larger MRSECs already do, they would be required to have a strong research program, ideally a cluster of at least three IRGs. Education and industrial outreach would follow the recommendations made in Chapters 4 and 5 of this report. In addition to a mandatory Research Experiences for Undergraduates (REU) program, MCEs would be encouraged to provide research opportunities to others, including faculty and students at primarily undergraduate and minority-serving institutions. A stronger national network of these sites should be established.

The proposed award period for the MCEs would be 6 years, as for MRSECs now, and there would be no limit to the overall lifetime. According to the sample budget outlined in Table 6.2, the committee believes that the three-IRG MCEs should have a target annual budget of $4.5 million. Considering the current research and infrastructure portfolio of the present MRSEC program, 10 or 11 MCEs would form a critical mass.

Materials Research Groups

A second theme of the findings section is that more heterogeneity and flexibility are needed in the types of research groups that should be deployed in materials research. The committee proposes that approximately half of the budget be used to fund collaborative research groups, similar in size and scale to the current IRGs. These groups could be called Materials Research Groups (MRGs) and would consist of three to seven PIs, with a student or postdoctoral associate per PI. The committee's example MRG budget of $600,000 is shown in Table 6.3. This funding mechanism differs from existing group research mechanisms in that the grant size would be larger than that for a focused research group, and the grant period would be 5 years, with no limits on renewals. This would allow MRGs to tackle substantive and long-term problems, to have continuity over the cycle of students, and to keep the new reviewing demands to a sustainable level.

The intention of the MRG part of the program is to diversify the research topic portfolio, increase timely response to new research opportunities, and provide institutions and individuals maximum flexibility in assembling the right team for the problem at hand. MRGs could have investigators from the same institution or from different institutions and could consist of any mix of disciplines appropriate to the research.

The committee notes that DMR currently has several mechanisms for supporting collaborative, group-based research at a level between that of individual investigators and a MRSEC.[5] These include Focused Research Groups (FRGs), Nanoscale Interdisciplinary Research Teams (NIRTs), and modest participation in medium-sized Information Technology Research (ITR) awards. The FRG program is an unsolicited program, similar to individual-investigator programs; these awards involve three or more faculty-level investigators with complementary expertise, the award size is on the order of $250,000 per year or greater, and the activities integrate research and education. Partnerships with industry and other sectors are encouraged. In 2006, DMR supported 33 FRGs, representing an annual investment of nearly $11 million. In 2005, DMR supported 36 active NIRT awards, although this program is being phased out; generally speaking, NIRTs acted as mini-centers and pursued a broad range of responsibilities similar to those of MRSECs. The committee therefore distinguishes its proposed MRG funding mechanism from the existing FRG program in three critical ways.

- FRG awards are typically for 3 to 4 years. An MRG award would be for 5 to 6 years, enabling a longer-term, and potentially more innovative, investigation. MRGs would be able to encourage "collaboration in conception" of research in addition to "collaboration in execution."
- The FRG program is not managed as a distinct budget element of DMR; the proposed MRG awards would be part of the joint solicitation with MCEs.
- MRGs would represent a key element of the revised MRSEC program portfolio. Competition for MRG awards would directly compare the research of MRGs and that of the MCEs.

The MRGs of the committee's proposal presumably would provide materials departments with the capability of responding more rapidly to developing opportunities. Each MRG would be focused on a general topic, somewhat similar to the IRGs of today's MRSECs. An MRG could make use of facilities, industrial partnerships, and education outreach resources facilitated by an MCE. However, the competitive review basis of the MRGs would focus on the research agenda. MRGs would even mimic some of the roles of the seed program that the present MRSECs use to invest a limited amount of resources in innovative topics that arise between competitive reviews.

In order for the committee's proposal to be successful (and to represent a step forward from the situation in the early 1990s with MRLs and MRGs), the program

[5]According to the NSF grants program guide, "A group proposal is one submitted by 3 or more investigators whose separate but related activities are combined into one administrative unit. A collaborative proposal is one in which investigators from two or more organizations wish to collaborate on a unified research project." Available at http://www.nsf.gov/funding/preparing/faq/faq_g.jsp?org=DMR#group.

of MCEs and MRGs must include a unified mechanism for review (both for entry to the program and for renewal). That is, the research elements of the MRGs and MCEs would have to compete directly with one another. This feature is critical in order to allow a level playing field between institutions with centers and those without: funding for research groups should be awarded to the most competitive proposals on the basis of the science alone. Additional resources for facilities and outreach would be awarded through a parallel but separate review process. A potential option to help reduce the load on the peer review community would be for NSF to offer merit-based opportunities to MRGs to renew rather than needing to recompete for the next cycle of support. The trade-off here would be in allowing MRGs a better chance at persistence and requiring NSF program managers to handle the additional workload of organizing and facilitating a full and complete open competition.

The committee's proposal is framed as a transformation of the MRSEC program that would initially be revenue-neutral, converting the roughly $50 million program into a 50/50 split of MCEs and MRGs. It is the committee's view that this two-pronged solution to the need for materials research centers would greatly enable future growth.

Operating as a Whole

No two MRSECs are the same, which argues for viewing the MRSEC portfolio as a whole, ensuring that all aspects of research and education are addressed, but not necessarily by each individual MRSEC. NSF encourages MRSECs to operate as a national network so that not only is each MRSEC greater than the sum of its parts, but also the full program is greater than the sum of its individual centers. Although there have been some efforts in this direction, the committee did not observe strong cooperation among the discrete centers of the program. Accomplishing this approach is essential for making the most efficient use of funding. The approach argues for flexibility in the degree of importance placed on the nonresearch aspects of the MRSEC—for example, education and industrial outreach, the details of the shared facilities, and so on. The committee notes that, viewing the MRSECs as a system:

1. No MRSEC has every component;
2. There is a geographic distribution, including regional availability of particular types of instruments and so on; this arrangement could also allow for a more rational approach to targeting underrepresented minorities in outreach programs, since these populations are not uniformly distributed throughout the country; and
3. There is appropriate sharing among MRSECs of lessons learned and other

opportunities, as well as leveraging of capabilities, and maybe even some staff for activities such as outreach.

Conclusion: NSF encourages MRSECs to operate as a national network. Although some efforts have been made in that direction, the committee did not observe strong cooperation among the discrete centers of the program. The MRSEC program is thus missing a clear opportunity to leverage resources and thereby strengthen the materials research enterprise as a whole.

The committee believes that in spite of the competition among centers for success in each cycle of the program competition and in spite of the 6-year time horizon for any one center's planning, substantial opportunities for synergy exist. Developing a hub-and-spoke model for promoting and sharing access to experimental facilities is one such avenue.[6] Furthermore, the opportunities for national networks for education and outreach and industrial collaboration are significant. Moreover, nationally coordinating these efforts of the individual MRSECs might help better define the objectives and procedures for these program elements. For instance, a shared database of effective EO activities and assessment tools would substantially assist new centers in implementing a meaningful program in education and outreach. The committee is not envisioning a whole-scale integration of every center into a consolidated entity but rather improved communication and coordination among them. Modest supplemental grants could assist in organizing joint workshops and enhancing access to industrial partners or shared facilities.

Recommendation: NSF should enable its materials research centers to play a greater role in advancing materials research.

As centers for teams of investigators, MRSECs could play a natural role in facilitating community formulation of initiatives in materials research. Such activities might include but not be limited to the following: organizing conferences and workshops addressing significant questions in materials research; creating and maintaining a national directory of MRSEC expertise and facilities; leveraging economies of scale in industrial and/or educational outreach; and providing geographically based infrastructure for materials research facilities. The committee notes, however, that this suggested direction for the MRSEC program should not be construed as yet another requirement for the centers. Rather, is it an affirmation of several grass-roots initiatives that have recently taken shape.

[6]See National Research Council, *Midsize Facilities: The Infrastructure of Materials Research*, Washington, D.C.: The National Academies Press, 2006, for details.

OUTLOOK

Interdisciplinary and multidisciplinary research will continue to be a hallmark of materials research, and NSF needs to continue to maintain a leadership role in supporting such activity. This committee endorses the concepts embedded in the current MRSEC program, but it encourages a significant realignment of budget, program structure, and management oversight to ensure optimum effectiveness of the NSF group research program in the face of limited resources.

Appendixes

A

Charge to the Committee

The purpose of this study is to:

1. Assess the performance and impact of the National Science Foundation's Materials Research Science and Engineering Centers program (MRSEC program); and
2. On the basis of current trends and needs in materials and condensed matter research, recommend future directions and roles for the program.

B

Meeting Agendas

FIRST MEETING
WASHINGTON, D.C.
NOVEMBER 18-19, 2005

Friday, November 18, 2005

Closed Session

1:00 p.m.	Welcome —M. Tirrell, Chair, MRSEC Impact Assessment Committee
1:05	Composition and Balance Discussion —D. Shapero, Director, Board on Physics and Astronomy (BPA)
2:00	Introduction to the National Research Council (NRC) —T.I. Meyer, BPA staff
2:30	General Discussion
2:45	Break

Open Session

3:00	General Discussion of the Study
4:00	Perspectives from the National Science Foundation (NSF) —U. Strom, NSF Materials Research Science and Engineering Centers Program Manager

4:30	Discussion
5:00	Perspectives from the MRSEC Directors Group —T. Russell, University of Massachusetts; Chair, MRSEC Directors Group
5:30	Discussion
6:00	Adjourn

Saturday, November 19, 2005

Closed Session

8:30 a.m.	Discussion
10:15	Break
10:45	Discussion of Assessment Strategies
Noon	Lunch
1:00 p.m.	Discussions, Including Work Plan
3:00	Adjourn

SECOND MEETING
SANTA BARBARA, CALIFORNIA
MARCH 8-9, 2006

Wednesday, March 8, 2006

Closed Session

8:30 a.m.	Welcome and Plans for the Meeting —M. Tirrell
9:00	Initial Findings of the Management and Facilities Group —F. DiSalvo, Cornell University
9:30	Initial Findings of the Industrial Collaboration Group —D. Dimos, Sandia National Laboratories
10:00	Initial Findings of the Research Group —P. Chaikin, New York University
10:30	Initial Findings of the Education and Outreach Group —D. Leslie-Pelecky, University of Nebraska
11:00	Committee Discussion

Open Session

Noon	Lunch

1:00 p.m.	Single-Investigator Research Perspectives —J. Brauman, Stanford University
1:30	Discussion
1:45	The Role of Industry in University-Based Center Research —C. Duke, Xerox (retired) [via teleconference]
2:15	Discussion
2:30	International Collaboration in Materials Research —A. Cheetham, University of California at Santa Barbara
3:00	Discussion
3:15	Break
3:30	Perspectives on Education and Outreach —P. Dixon, National High Magnetic Field Laboratory [via teleconference]
4:00	Discussion
4:15	Instrumentation and Facilities —R. Nuzzo, University of Illinois at Urbana-Champaign
4:45	Discussion

Closed Session

5:00	Committee Discussion
5:45	Adjourn

Thursday, March 9, 2006

Closed Session

8:00 a.m.	Discussions
8:30	Working Group Breakout Sessions
10:30	Break
10:45	Group Discussion
11:45	Plans for Next Meeting
Noon	Adjourn

Open Session

1:00 p.m.	Tour of Materials Research Laboratory —G. Fredrickson, M. Evans, University of California at Santa Barbara
4:00	Adjourn

THIRD MEETING
WASHINGTON, D.C.
JUNE 12-13, 2006

Monday, June 12, 2006

Closed Session

8:30 a.m.	Welcome and Plans for the Meeting —M. Tirrell
9:00	Update from Industrial Collaboration Group —D. Dimos
9:30	Update from Research Group —R. Nuzzo
10:00	Update from Management and Facilities Group —F. DiSalvo
10:30	Update from Education and Outreach Group —D. Leslie-Pelecky
11:00	Committee Discussion
Noon	Lunch
1:00 p.m.	Themes Characterizing Impact of the MRSECs —M. Tirrell
2:00	Breakout Sessions
3:30	Break
3:45	Reconvene for Group Discussion
5:15	Adjourn

Tuesday, June 13, 2006

Closed Session

8:00 a.m.	Discussion of Findings and Recommendations
10:30	Break
10:45	Review of Plans for Site Visits —T.I. Meyer
11:30	Plans Going Forward —M. Tirrell
Noon	Adjourn

FOURTH MEETING
IRVINE, CALIFORNIA
SEPTEMBER 19-20, 2006

Tuesday, September 19, 2006

Closed Session

8:30 a.m.	Welcome and Plans for the Meeting —M. Tirrell
8:35	Review of Progress: What Have We Learned?
8:45	Mixed Breakout Sessions
11:30	Reconvene and Committee Discussion
Noon	Working Lunch
1:00 p.m.	Findings and Recommendations
3:30	Break
3:45	Continued Discussion
5:30	Adjourn

Wednesday, September 20, 2006

Closed Session

8:30 a.m.	Plans for the Day —M. Tirrell
8:45	Breakout Sessions
10:00	Break
10:15	Group Discussions
11:45	Plans Going Forward —M. Tirrell
Noon	Adjourn

C

List of Current Research Topics of MRSEC Interdisciplinary Research Groups

The specific research programs at each Materials Research Science and Engineering Center (MRSEC) are determined by the topics of each Interdisciplinary Research Group (IRG). The list below of 2006 IRG research topics is from the Web site http://www.mrsec.org.

BIOMOLECULAR/BIOMIMETIC MATERIALS

- Patterns, Gradients and Signals in Soft Biomaterials—California Institute of Technology
- Engineering Materials and Techniques for Biological Studies at Cellular Scales—Harvard University
- Materials and Physiology—Harvard University
- Specific, Reversible and Programmable Bonding in Supra- and Macromolecular Materials—University of California at Santa Barbara
- Molecular and Nanoscale Motors—Pennsylvania State University
- Design and Synthesis of Response Driven Macromolecules—University of Southern Mississippi
- Template Synthesis of Nanowire/Nanotube Heterostructures—University of Maryland
- Bio-Interfacial Science—University of Chicago
- Designed Programmable Membranes—University of Pennsylvania
- Filamentous Networks and Structured Gels—University of Pennsylvania

- De Novo Synthetic Protein Modules for Light-Capture and Catalysis—University of Pennsylvania
- Biological Synthesis and Assembly of Macromolecular Materials—California Institute of Technology

COATINGS/CERAMICS

- Synergistic Linear and Nonlinear Phenomena in Multifunctional Oxide Ceramic Systems—Northwestern University
- Responsive Films and Film Formation—University of Southern Mississippi
- Oxide-Based Hierarchical Interfacial Materials—University of Pennsylvania
- Biological Synthesis & Assembly of Macromolecular Materials—California Institute of Technology

CONDENSED MATTER PHENOMENA

- Electronic Interfaces—Cornell University
- Nanoscale Growth—Cornell University
- Nanomechanics—Cornell University
- Electrons in Confined Geometries—Pennsylvania State University
- Spin and Spin Coherence Dynamics of Tunable Electrochemically Synthesized Nanostructures—University of Maryland
- Interplay of Magnetism and Transport in Correlated Electronic Materials—Princeton University
- Microphotonic Materials and Structures—Massachusetts Institute of Technology
- Fluid Flows-Singularities and Microscales—University of Chicago
- Jamming, Slow Relaxation and Rigidity Onset in Materials Far from Equilibrium—University of Chicago

MAGNETICS/FERROELECTRICS/SPINTRONICS

- Spin and Charge Quantum Transport in Organic/Magnetic Heterostructures for Spintronics and Optoelectronics—California Institute of Technology
- Ferroelectric Photonic Materials—California Institute of Technology
- Science and Engineering of Magnetoelectronics—Johns Hopkins University
- IRG—Yale University
- Electronic Interfaces—Cornell University
- Nanoscale Growth—Cornell University

- Magnetic Heterostructures—University of Minnesota
- Nanomagnetism: Fundamental Interactions and Applications—University of Nebraska
- Spin Polarization and Transmission at Nanocontacts and Interfaces—University of Nebraska
- Electrons in Confined Geometries—Pennsylvania State University
- Multifunctional Magnetic Oxides—University of Maryland
- Spin and Spin Coherence Dynamics of Tunable Electrochemically Synthesized Nanostructures—University of Maryland
- Interplay of Magnetism and Transport in Correlated Electronic Materials—Princeton University
- Electronic Transport in Mesoscopic Semiconductor and Magnetic Structures—Massachusetts Institute of Technology
- Oxide-Based Hierarchical Interfacial Materials—University of Pennsylvania
- Dynamics and Transport in Nanostructured Magnetic Materials—University of Alabama

MECHANICS OF MATERIALS

- Tailored Interfaces—University of Massachusetts at Amherst
- Mechanics of Amorphous and Nanoscale Metal Composites and Foam Structures—California Institute of Technology
- Center Research Summary—Cornell University
- Advances in Continuum Simulation Methods—University of Virginia
- Multiscale Mechanics of Films and Interfaces—Harvard University
- Engineering Materials and Techniques for Biological Studies at Cellular Scales—Harvard University
- Nanomechanics—Cornell University
- Adhesion, Deformation and Transport at Contacts in Small Structures—Princeton University
- Science and Engineering of Solid-State Portable Power Structures—Massachusetts Institute of Technology
- Mesoscale Interface Mapping Project—Carnegie Mellon University

NANOSTRUCTURES/NANOPARTICLES

- Mechanics of Amorphous and Nanoscale Metal Composites and Foam Structures—California Institute of Technology
- Nanostructures-Growth and Characterization—University of Oklahoma/University of Arkansas

- Micro- and Nano-Mechanics of Electronic and Structural Materials Research—Brown University
- Novel Processing Methods for Nanostructured Polymer Blends and Composites—Northwestern University
- Plasmonics and Molecular Based Electronics: Fundamentals and New Tools—Northwestern University
- Self-organization in the Synthesis of Nanostructured Materials—Northwestern University
- Elucidation of Fundamental Nucleation Localization Mechanisms—University of Virginia
- Nanoscale Surface Modification by the Focused Ion Beam—University of Virginia
- Electronic Interfaces—Cornell University
- Photonic Particles—Cornell University
- Nanoscale Growth—Cornell University
- Directed Nano-assemblies and Interfaces for Advanced Electronics—Stanford/IBM ARC/UC Davis/UC Berkeley
- Nanoparticle-Based Materials—University of Minnesota
- Nanomechanics—Cornell University
- Nanostructured Materials by Molecular Beam Epitaxy—University of California at Santa Barbara
- Electrons in Confined Geometries—Pennsylvania State University
- Spin and Spin Coherence Dynamics of Tunable Electrochemically Synthesized Nanostructures—University of Maryland
- Template Synthesis of Nanowire/Nanotube Heterostructures—University of Maryland
- Diffusion and Wettability in Porous Nanoparticles—University of Maryland
- Guided Self-Assembly—Princeton University
- Adhesion, Deformation and Transport at Contacts in Small Structures—Princeton University
- Electronic Transport in Mesoscopic Semiconductor and Magnetic Structures—Massachusetts Institute of Technology
- Nanostructured Polymer Assemblies—Massachusetts Institute of Technology
- Hierarchically Assembled Molecular and Hybrid Organic-Inorganic Materials—University of Chicago
- Structural Integrated Films Containing Nanoparticles—Columbia University
- Materials for Information Storage—University of Alabama

POLYMERS

- Seed Project: Methanol Generation and Its Efficient Use in Fuel Cells—California Institute of Technology
- Research Groups Overview—SUNY at Stony Brook
- IRG #2—Novel Processing Methods for Nanostructured Polymer Blends and Composites—Northwestern University
- Center Research Summary—Cornell University
- IRG1: Microstructured Polymers—University of Minnesota
- Soft Cellular Materials—University of California at Santa Barbara
- Template Synthesis of Nanowire/Nanotube Heterostructures—University of Maryland
- Guided Self-Assembly—Princeton University
- Nanostructured Polymer Assemblies—Massachusetts Institute of Technology
- Hierarchically Assembled Molecular and Hybrid Organic-Inorganic Materials—University of Chicago
- Functional Cylindrical Assemblies—University of Pennsylvania
- Structured Materials in Supercritical Fluids—University of Massachusetts at Amherst
- Aqueous Polymer Assembly—University of Massachusetts at Amherst

SEMICONDUCTORS/PHOTONICS/ORGANIC ELECTRONICS

- Plasmonics and Molecular Based Electronics: Fundamentals and New Tools—Northwestern University
- Controlling Interfaces in Semiconductor Nanowires—Northwestern University
- Electronic Interfaces—Cornell University
- Photonic Particles—Cornell University
- Nanoscale Growth—Cornell University
- Crystalline Organic Semiconductors—University of Minnesota
- Optical Metamaterials—Pennsylvania State University
- Oxides as Semiconductors—University of California at Santa Barbara
- Low Dimensional Interfaces—University of Maryland
- Interplay of Magnetism and Transport in Correlated Electronic Materials—Princeton University
- Adhesion, Deformation and Transport at Contacts in Small Structures—Princeton University

SOFT MATERIALS, COLLOIDS

- Patterns, Gradients and Signals in Soft Biomaterials—California Institute of Technology
- Single Interdisciplinary Research Group—University of Colorado
- Center Research Summary—Cornell University
- Interface-Mediated Assembly of Soft Materials—Harvard University
- Molecular and Nanoscale Motors—Pennsylvania State University
- Template Synthesis of Nanowire/Nanotube Heterostructures—University of Maryland
- Guided Self-Assembly—Princeton University
- Jamming, Slow Relaxation and Rigidity Onset in Materials Far from Equilibrium—University of Chicago
- Functional Cylindrical Assemblies—University of Pennsylvania
- Filamentous Networks and Structured Gels—University of Pennsylvania

SYNTHESIS/PROCESSING

- The Synthesis of Deuterated-Rhodamine 6G—Northwestern University
- Photonic Particles—Cornell University
- Nanoscale Growth—Cornell University
- Nanomechanics—Cornell University
- Chemically Advanced Nanolithography—Pennsylvania State University
- Guided Self-Assembly—Princeton University

D

Further Information on Education and Outreach Activities

BRINGING DISCUSSIONS ABOUT THE SOCIAL RAMIFICATIONS OF TECHNOLOGY INTO THE CLASSROOM

The MRSEC at the University of Wisconsin-Madison, partnering with another center on campus, helped to create a new Science and Technology Studies course, "Nanotechnology and Society."[1] This course introduces undergraduate students to the necessity of thinking about how technology influences society through several broad objectives. These include introducing the nanoscale science field, considering the social ramifications of nanotechnology, and developing analytical and communication skills. Participation in class activities in the discussion-oriented class is essential. Activities include student-led group discussions, class presentations, and group tasks. Students also complete several essays, two exams, and an individual research project in which each student becomes a class expert on a selected topic and gives reports on progress as would a real-world research group. This Science and Technology Studies course does require some basic science education, which is covered early in the semester.

During the semester, assessment surveys are completed to evaluate the students' progress and to provide feedback. Survey results from the first semester show that as the course progressed, the students demonstrated a growing understanding of

[1]See C. Tahan, R. Leung, G.M. Zenner, K.D. Ellison, W.C. Crone, and C.A. Miller, "Nanotechnology and Society: A Discussion-Based Undergraduate Course," *American Journal of Physics* 74 (5): 4-11 (2006).

how society will be affected by nanotechnologies, and how society can in turn affect the course of technological advancement and application. When the semester was over, the students were able to frame pertinent questions about the implications of nanoscale science and engineering. Most said they were very well prepared to explain the concepts of nanoscale science and engineering. Although the course did not encourage the students to follow a career in policy or science and technology studies, they all felt the course was worthwhile. Many in the class said that their perspective on science, technology, and societal implications had changed from a belief that all technological advances are a good thing to a more general acknowledgment and understanding of the social issues behind new advancements.

The University of Wisconsin-Madison case is a clear example of how the MRSEC program has a positive impact on undergraduate learning. Scientists, technologists, and students need to consider the effects of technology on society, and it is imperative that educators join together to involve their undergraduate students. Through courses that introduce a new field like nanotechnology, students receive a foundation that is necessary for understanding the issues of technological change and development. Efforts such as this, made possible in part by the MRSEC program, are a true innovation in science education.

MRSEC EDUCATION AND OUTREACH MEETINGS

- *October 21-23, 1998, University of California, Santa Barbara:* The "Making Connections" workshop had more than 75 participants, including MRSEC directors and outreach coordinators, university science faculty, high school and community college teachers, and students. Participants summarized current issues in science education, including presenting science to the wider community, engaging student interest in investigation, building partnerships with K-12 schools, creating resources for educational outreach and program evaluation.
- *November 13-14, 2003, University of Virginia:* One-day symposium for education and outreach (EO) directors to make short presentations of their work. Twenty-six EO coordinators and 27 center directors attended this meeting. The University of Virginia made a compilation of the programs and achievements of each MRSEC for 2003-2004 that offers a single-page synopsis highlighting examples of EO highlights.[2]
- *April 13-15, 2006, University of Chicago:* A meeting of MRSEC and EO directors, with a topical focus on evaluation and assessment of educational programs.

[2]See http://www.mrsec.virginia.edu/nugget5emed.htm.

- *Other meetings:* MRSEC EO coordinators have been involved in other meetings.
- *The Research Experiences for Teachers (RET) network*[3] unites those who run and evaluate RET programs; many MRSECs are active in this group. RET conferences were held in 2002 and 2003. Sessions at meetings were sponsored in 2004 (American Chemical Society meeting) and 2006 (National Science Teachers Association regional meeting). In addition to conferences, the RET network Web site has a collection of assessment tools, including pre- and post-program survey forms.
- *National Research Centers Educators Network:* A group of EO coordinators from NSF centers, including MRSECs, Science and Technology Centers, and Engineering Research Centers has formed the National Research Centers Educators Network (NRCEN).[4] The goals of the group are to identify and disseminate models, tools, resources, experiences; determine mechanisms or strategies to enhance centers' efforts; and identify and address priority issues specific to centers. Meetings have been held in 2001 (Cornell University), 2002 (University of California at Santa Cruz), 2004 (University of Florida), 2005 (California Institute of Technology), and 2007 (University of Michigan).
- *Fall 2004 Materials Research Society Meeting:* A group of EO coordinators obtained NSF funding to bring RET teachers from MRSECs to the 2004 Fall Materials Research Society (MRS) meeting. The teachers attended the MRS education symposia and participated in hands-on workshops about MRSEC-related curricular materials.

[3]See http://www.retnetwork.org/.
[4]See http://www.nrcen.org/.

E

Selected Acronyms

AFOSR	Air Force Office of Scientific Research
ARO	Army Research Office
ARPA	Advanced Research Projects Agency
CHESS	Cornell High Energy Synchrotron Source
COSEPUP	Committee on Science, Engineering, and Public Policy
DARPA	Defense Advanced Research Projects Agency
DMR	Division of Materials Research (National Science Foundation)
DOD	Department of Defense
DOE	Department of Energy
DURIP	Defense University Research Instrumentation Program
EO	education and outreach
ERC	Engineering Research Center
FRG	Focused Research Group
IDL	Interdisciplinary Laboratory
IGERT	Integrative Graduate Education and Research Traineeship
IMR	Instruments for Materials Research
IP	intellectual property
IRG	Interdisciplinary Research Group
ITR	Information Technology Research
MCE	Materials Center of Excellence
MPI	Max Planck Institute

MPS	Mathematics and Physical Sciences (National Science Foundation)
MRG	Materials Research Group
MRI	Major Research Instrumentation
MRL	Materials Research Laboratory
MRSEC	Materials Research Science and Engineering Center
MRSEC program	Materials Research Science and Engineering Centers program
MSE	materials science and engineering
MSI	minority-serving institution
MURI	Multidisciplinary University Research Initiative
NASA	National Aeronautics and Space Administration
NIRT	Nanoscale Interdisciplinary Research Team
NIST	National Institute of Standards and Technology
NRC	National Research Council
NRCEN	National Research Center Educators Network
NSEC	Nanoscale Science and Engineering Center
NSF	National Science Foundation
NSTC	National Science and Technology Council
NUE	Nanoscale Undergraduate Education
OMB	Office of Management and Budget
ONR	Office of Naval Research
PI	principal investigator
PREM	Partnerships for Research and Education in Materials
RET	Research Experiences for Teachers
REU	Research Experiences for Undergraduates
RFP	Request for Proposal
R&RA	research and related activities
SEF	shared experimental facilities
STC	Science and Technology Center
STEM	science, technology, engineering, and mathematics
TEM	tunneling electron microscope
UARC	University-Affiliated Research Center
URM	underrepresented minorities

F

Data-Gathering Tools

The MRSEC Impact Assessment Committee conducted numerous data-gathering activities in order to be able to circumscribe the current level of effort in the MRSEC program accurately. Owing to the diverse nature of the Materials Research Science and Engineering Centers (MRSECs) and the program's numerous requirements, it was necessary to employ multiple approaches to obtain the best (and most) data possible. In addition to requesting the most recent and very first annual reports from each MRSEC—responses were received from 27 of 29 and 25 of 29 MRSECs, respectively—the committee developed and used the data-gathering tools presented in this appendix to conduct its study of the MRSEC program. After receiving the data on a particular request, the committee members and staff compiled the data into a summary form and discussed them at length. As the data suggested particular lines of inquiry, the committee followed up with subsequent data-gathering efforts.

DATA REQUEST TO MRSEC DIRECTORS

The committee sent a questionnaire to all 29 MRSECs, addressed to each center's director (see Box F.1). The topics covered the MRSECs' perceived scientific accomplishment, student output, education and outreach, industrial collaborations, and facilities and instrumentation. The committee received full responses from 23 of 29 MRSECs.

BOX F.1
Information Request to Center Directors for NRC MRSEC Impact Assessment Committee

*Please address these questions first and return this form to the National Research Council by Friday, February 24, 2006. *If you would, please send any evaluations as requested in number 2 below as soon as possible.*

Name of Center: ______________________________

1. For the following, please indicate what you believe to be your MRSEC's top 5:
   a. scientific questions currently addressed.
   b. lifetime accomplishments.
   c. most highly-cited papers. Please list full citation information.
   d. most important contributions to materials research science and engineering.
   e. most successful students who have gone on to careers in academe or industrial research. Please also indicate their key contributions.

2. If your center has engaged in or commissioned any evaluations of the education and public outreach component, please briefly describe them and attach a copy of the evaluation report. (*Please see above—the committee is quite eager to learn from you!)

3. What Shared Experimental Facilities have been established at your center? What research have they enabled for the MRSEC and beyond?

4. What are the goals of your industrial collaborations?

5. What do you feel would be the optimal outcome of your MRSEC's industrial collaboration effort if the interaction were as successful as possible?

6. What are the education/outreach goals of your MRSEC? How were the education/outreach goals of your MRSEC determined?

7. How does materials research conducted at your center differ from that typical of single investigators at your institution? What is your impression of the reason for this difference?

8. If you could propose one change to improve the NSF MRSEC program, what would it be and why?

DATA REQUEST TO NSF MRSEC PROGRAM MANAGERS

The committee sent a formal data request to the NSF MRSEC Program Managers (see Box F.2). NSF responded fully to all requests except for request 8(c), for which data were incomplete.

BOX F.2
Information Request to NSF MRSEC Program Managers for NRC MRSEC Impact Assessment Committee

1. Please provide copies of the reports of external committees of visitors for NSF/DMR over the past 10-12 years.

2. Please provide copies of external review reports for individual MRSECs over the past 5 years.

3. Please provide a breakdown of budget information in as-spent dollars for each year for the past 15 years for the following categories:
   a. Total MRSEC program budget (if there is a standard, few-category breakdown, please provide it as well)
   b. Total NSF/DMR budget
   c. Fraction of DMR budget spent altogether on centers

4. If possible, please provide a year-by-year total budget for the former MRL program along with any appropriate and reliable breakdown into categories.

5. Please provide the names of the 30 institutions that receive the most NSF/DMR funding in FY2005 (the Tom Weber exercise).

6. Please provide contact info for NSF/DMR counterparts in other countries such as Japan, China, Korea, Germany, France, Netherlands, United Kingdom, and so on. Please use your best judgment!

7. Please provide year-by-year totals for numbers of patents filed under the MRSEC program for the past 10-12 years.

8. Please provide year-by-year totals for the following "head count" metrics, including a description of what is tallied for each metric:
   a. Number of (graduate) students at MRSECs and the number of (graduate) students supported by DMR;
   b. Number of postdoctoral researchers at MRSECs and the number of postdoctoral researchers supported by DMR; and
   c. As available, please also provide year-by-year totals of the number of students who moved on to jobs in academia, industry, or elsewhere.

9. Please provide a copy of the guidelines for MRSEC annual reports. If the guidelines changed significantly over the course of the program, please include a copy of the oldest guidelines as well.

10. Please provide a copy of any past reports that have reviewed the MRSEC program.

DATA REQUEST REGARDING MRSEC EDUCATION AND OUTREACH

The committee sent a data request to the MRSEC directors and education and outreach (EO) coordinators (if applicable) seeking to understand the breadth of EO activities conducted and the mechanism by which MRSECs fund them (see Box F.3). The chart, which was quite instructive to the committee, helped unravel the complex nature of these programs. The committee received 15 of 29 responses for this data request.

BOX F.3
Information Request Regarding Education and Outreach for NRC MRSEC Impact Assessment Committee

Activity	Have done previously	Are currently doing	Breakdown of total support for activity (approx percent)[1]	MRSEC support for activity (approx percent)[2]	Other sources of support	Approx # of MRSEC Researchers involved per year
Research Experiences for Teachers (RET)						
Research Experiences for Undergraduates (REU)						
Other K-12 Teacher Professional Development (including workshops, but not REU)						
K-12 curriculum development / enhancement						
Undergraduate curriculum development / enhancement						
Graduate student curriculum development / enhancement						
Public Outreach (science museum exhibits, talks for the general public)						
Other (please describe below)						
Additional activity						
Additional activity						
Additional activity						

[1]Given the entire budget for an activity, what percentage is supported by the MRSEC grant?
[2]What percentage of the entire MRSEC grant is spent on this activity?

SITE VISITS

As described in Box 3.2 in Chapter 3 of this report, the committee conducted a series of site visits at institutions that either have (or had) a MRSEC or a similar center-based research structure. These site visits consisted of speaking with center leadership, research faculty, students, education and outreach coordinators, and industrial collaboration coordinators, in addition to departmental and university leadership. The committee visited the following institutions:

- Boston University:
  —Center for Nanoscience and Nanobiotechnology
  —Center for Subsurface Sensing and Imaging Systems
  —Center for Information Systems and Engineering
- California Institute of Technology: Center for the Science and Engineering of Materials (MRSEC)
- Harvard University: MRSEC
- Massachusetts Institute of Technology: Center for Materials Science and Engineering (MRSEC)
- Michigan State University: Center for Sensor Materials (MRSEC, now closed)
- University of California at San Diego: Center for Magnetic Recording Research
- University of California at Santa Barbara: Materials Research Laboratory
- University of Florida:
  —Microkelvin Laboratory
  —Nanoscience Institute for Medical and Engineering Technology
  —Major Analytical Instrumentation Center
  —Center for Condensed Matter Sciences
  —Center for Nano-Bio Sensors
  —Particle Engineering Research Center (ERC)
  —Center for Macromolecular Science and Engineering
  —Quantum Theory Project
  —Center for Precollegiate Education and Training
  —South East Alliance for Graduate Education and the Professoriate
- University of Michigan: Engineering Research Center for Reconfigurable Manufacturing Systems (ERC)
- University of Southern California: Biomimetic Microelectronic Systems (ERC)
- University of Southern Mississippi: Center for Response-Driven Polymeric Films (MRSEC)

The committee used a standardized set of questions during the site visits in order to be able to easily compare responses (see Box F.4). Since site visits included several centers outside the MRSEC program, the committee made small adjustments to the document as appropriate.

BOX F.4
Questions for Site Visits

A. PURPOSES OF THE MRSEC PROGRAM

Why should a MRSEC-like program continue as a mode of support at NSF? Why not just have individual investigator grants? The point of this discussion is to determine to what extent the original goals and intentions of the centers have been achieved AND to determine if centers, perhaps in a new mode, are still appropriate or necessary for the future of materials research. We will need as much quantitative data as possible, but also some qualitative information.

1. What is different about the quality or character of MRSEC research relative to single investigator research at your institution? To the extent possible, provide data to support your contentions. Also, please provide a specific example of a research problem in your MRSEC that well exploits these unique characteristics.
2. Why not have individual investigator grants, and let groups "self-assemble" if they think it is important? Are there examples of such "self-assembly" at your institution? If so, how many people are/were involved? Are they interdisciplinary?
3. What is the business model for supporting the shared experimental facilities in your MRSEC? If there are other materials research facilities on campus, in general, how are they managed and supported?

B. EVOLUTION OF RESEARCH AND THE ROLE OF MRSECs

How and when does the science evolve or develop new themes? Does having a center lead to more or less agility in initiating or exploring new topics? Please identify examples. The point of this discussion is to explore the tension between providing continuing investment in topical areas of critical scientific import and in generating/exploring entirely new topics. There is no "right answer" here, but we need to understand how this tension is managed and why it is managed in the way it is.

1. What is the longevity of the different IRG topics in your MRSEC? How does this compare to the same for single investigator grants in the relevant departments?
2. If you have more than one IRG at your center, in what ways do they come into contact, in terms of science, shared facilities, or students? How and when do the research topics of IRGs change?
3. If your MRSEC supports Seeds or other "startup" ideas, how do they function? What is the typical period of support? When complete, what fraction continue in some way? Within the MRSEC? Outside the MRSEC? Do most new IRGs develop from seeds?
4. Are there other ways that new topics and ideas are introduced to your center? Is it easier to obtain support for new ideas elsewhere? What mechanisms could expand your center's capability to start work on new topics?

continued

BOX F.4
Continued

C. BUDGETS AND RESOURCES

The intent of this question is both historical and forward-looking. The budgets at NSF for the past 6 years were very constrained. This has led to a call by some to put a larger fraction (or 100%) of the DMR budget into single investigator grants. If MRSEC-like centers continue into the future, how can they be as effective as possible in their mission within the resource constraints?

1. The cost of supporting a student (tuition, stipend, fringe, overhead) or post-doctoral at most universities has increased at a rate higher than general U.S. inflation. Please provide the yearly costs per graduate students and post-doctoral researcher for participants in your MRSEC since its inception. What has been the average inflation rate for the last five years in those costs?
2. How do you manage the MRSEC budget under 6 years of flat funding (which is steadily eroded by inflation)? Have you eliminated functions or activities in the center as a result of flat funding?
3. What level of support is provided to the center by the university or any other source outside the NSF MRSEC program? In what form is that support made (e.g., cash, space, people, and so on)? Why is this support provided? Would similar support likely be forthcoming without an externally funded center? Please provide examples and counter-examples.
4. (If we have questions after reviewing the appendices from the annual reports.)
   a. Averaged over the PIs in your MRSEC, what fraction of their total research support comes from the MRSEC?
   b. Over the recent history of your center, what fractions of the budget have been devoted to the following: (a) research, (b) education and outreach, and (c) industrial outreach and collaboration? How and why should this relative balance change in the future?

D. FUTURE DIRECTIONS FOR THE PROGRAM

1. Given the resources afforded your center, what would be the ideal interdisciplinary research and education center for your institution? Key components of the program?
2. What are "Grand Challenge" topics in materials (science, engineering, technology)? Of these, which do you foresee will require center-like approaches to explore and develop in a timely manner? Some

view many IRG topics in different centers as rather similar, even duplicative. How can the entire MRSEC program support a broader scope of research?

3. What broader impacts has your center had on the materials research effort at your university? How have you demonstrated success in any of these areas?
4. Has there been a significant change in the materials research program due to the establishment of the MRSEC? (If an MRL predates the MRSEC, how did it shape on-campus culture?)
5. How do you judge success in industrial outreach & knowledge transfer? What criteria and/or metrics are appropriate to evaluating progress?
6. How is the educational experience (for graduate and undergraduate students) enhanced by being part of MRSEC sponsored industrial outreach? Can you contrast to the experience for students not involved in MRSEC industrial outreach?
7. What changes (if any) are needed in your industrial outreach and knowledge transfer effort to respond to changes in the evolving industrial climate?
8. How do researchers feel about the role of EO within their MRSEC program? How does participation in MRSECs EO activities affect researchers? (Ask of faculty and students.)
9. How are you planning on handling expanded mandates for assessment of your EO program?
10. Reflections on the review process:
    - What works well?
    - How would you improve the process?

E. MISCELLANEOUS

What did we miss? What else do you think is important or should be included in our report and recommendations? Our goal is to understand what (if any) differences exist for students' educational experience based on their involvement in the MRSEC program (or if the presence of a MRSEC on campus provides comparable benefit to all students doing materials research).

F. DISCUSSIONS WITH STUDENTS AND OTHER USERS

We would like to talk about some of the following topics with students and other participants in the MRSEC.

1. Compared to your peers, how do you perceive that your experience in the MRSEC is different?
2. What are your aspirations beyond graduate school?

G

Biographical Sketches of Committee Members and Staff

COMMITTEE MEMBERS

Matthew V. Tirrell, University of California at Santa Barbara (NAE), Chair

Dr. Tirrell is dean of engineering and a professor in the Chemical Engineering and Materials Departments at the University of California at Santa Barbara. His research interests are in the manipulation and measurement of interfacial properties of materials used in coatings, adhesion, lubrication, and bioengineering. Before Santa Barbara, he was head of chemical engineering and materials science at the University of Minnesota. Dr. Tirrell earned his Ph.D. from the University of Massachusetts in 1977. Among his many awards, he has received the Charles M.A. Stine Award of the American Institute of Chemical Engineers, the John H. Dillon Award of the American Physical Society (APS), and the Alumni Merit Award from Northwestern University. He has been elected to the National Academy of Engineering and has served on the National Research Council's (NRC's) Board on Chemical Sciences and Technology.

Kristi S. Anseth, University of Colorado at Boulder

Dr. Anseth is a professor of molecular biotechnology at the University of Colorado and an associate professor of surgery at the University of Colorado. Her research interests are in biomaterials, tissue engineering, and biomedical applications of degradable polymer networks. She has received the Alan T. Waterman Award from the National Science Foundation, the Outstanding Young Investigator Award from the Materials Research Society, and the Boulder Faculty Assembly Award for Excellence in Research, as well as the Scholarly and Creative Work Award.

Meigan Aronson, Brookhaven National Laboratory

Dr. Aronson is a research scientist at Brookhaven National Laboratory. She was most recently a professor of physics at the University of Michigan. She was also associate director of the Michigan Electron Microbeam Analysis Laboratory, a user facility for the university research community. Dr. Aronson graduated with a Ph.D. from the University of Illinois at Urbana-Champaign in 1988. Her research is on quantum-phase transitions, phase behaviors of low-density metals, and novel magnetism. The central focus of her research is the exploration of magnetism in metals and the properties of electron gas at low densities, where strong and unscreened Coulomb interactions are expected to lead to unusual types of charge and spin order, especially in very large magnetic fields. Her group uses neutron scattering, as well as a variety of transport, magnetic, and thermal measurements, to probe the ground state and its excitations at low temperatures and at high magnetic fields up to as large as 60 tesla, and at pressures as large as 200,000 atmospheres. Dr. Aronson is a fellow of the APS and recently served on the NRC's Committee on Opportunities in High Magnetic Field Science.

David M. Ceperley, University of Illinois at Urbana-Champaign

Dr. Ceperley is a professor of physics and a staff scientist at the National Center for Supercomputing Applications. He worked at both the Lawrence Berkeley National Laboratory (LBNL) and the Lawrence Livermore National Laboratory (LLNL) before coming to the the University of Illinois in 1987. His research interests include quantum Monte Carlo methods and quantum many-body systems, studying systems such as the energy of an electron gas, the electronic structure of condensed matter, and the macroscopic properties of liquid helium.

Paul M. Chaikin, New York University (NAS)

Dr. Chaikin is a professor of physics at New York University. His research interests include soft condensed-matter physics, colloids, nanolithography, and low-dimensional strongly correlated electron systems (especially organic superconductors) using high magnetic fields. Dr. Chaikin is a fellow of the American Academy of Arts and Sciences, a fellow of the American Physical Society, and a past winner of the prestigious Guggenheim Fellowship and A.P. Sloan Foundation Fellowship awards. He was elected to the National Academy of Sciences in 2004.

Ronald C. Davidson, Princeton University

Dr. Davidson is a professor of astrophysical sciences at Princeton University. His research interests are in pure and applied plasma physics, including non-neutral plasmas, nonlinear effects and anomalous transport, kinetic equilibrium and stability properties, and intense charged-particle beams. As an outsider to the NSF MRSEC program, he has deep knowledge of both the Department of Energy and large research centers. Dr. Davidson has served as the director of the Princeton Plasma Physics Laboratory; as the assistant director for Applied Plasma Physics Office of Fusion Energy, Department of Energy; as the director of the Massachusetts Institute of Technology's Plasma Fusion Center; as the first chair of the Department of Energy's Magnetic Fusion Advisory Committee; and as chair of the American Physical Society's Division of Plasma Physics. He has been an American Physical Society (APS) Councilor and a member of the APS Executive Board. Dr. Davidson has participated in numerous national and international committees on plasma physics, accelerator physics, and fusion research, including many review panels of the National Research Council.

Duane B. Dimos, Sandia National Laboratories

Dr. Dimos is deputy director of the Materials and Process Sciences Center at Sandia National Laboratories. His research has focused on thick- and thin-film electronic ceramics, and for many years he led work to develop ferroelectric thin films for a variety of applications. In addition, he has done fundamental work on superconducting thin films and diffusion and defect processes in mixed oxides.

Francis J. DiSalvo, Cornell University (NAS)

Dr. DiSalvo is professor of physical science at Cornell University in the Chemistry Department. His research interests are broadly in the synthesis and characterization of materials, recently focusing on the problem of fuel cells. Dr. DiSalvo was director of the Cornell Center for Materials Research, one of 29 such national centers supported by the National Science Foundation. He earned his Ph.D. in applied physics in 1971 from Stanford University. He then joined the research staff at AT&T Bell Laboratories (now Lucent Technologies), where he later headed several research departments. In 1986, Dr. DiSalvo moved to Cornell's Chemistry Department. His research interests are in the synthesis and characterization of inorganic compounds, and he is currently specializing in nitrides and intermetallic materials with novel crystal structures. Dr. DiSalvo is a fellow of the American Physical Society and of the American Association for the Advancement of Science. He has received the APS International New Materials Prize. He is also a member of the American Chemical Society, the Materials Research Society, and the National Academy of Sciences. Dr. DiSalvo is a past member of the NRC's National Materials Advisory Board. He was recently a member of the Solid State Sciences Committee's Committee on Smaller Facilities, which examined the issues of midsize facilities broadly within materials research.

Edith M. Flanigen, UOP, Inc. (retired) (NAE)

Dr. Flanigen is retired from UOP, Inc., where she was a leading researcher in materials synthesis, with an emphasis on petroleum refining methods and synthetic emeralds of high quality. She has served on the industrial review boards of several university centers.

Thomas F. Kuech, University of Wisconsin-Madison

Dr. Kuech is a professor in the Department of Chemical and Biological Engineering at the University of Wisconsin-Madison. His research interests are broadly in materials synthesis and processing, with an emphasis on semiconductor processing and electronic materials. He has chaired the Electronic Materials Conferences and is a fellow of the American Physical Society.

Diandra L. Leslie-Pelecky, University of Nebraska-Lincoln

Dr. Leslie-Pelecky is a professor of physics at the University of Nebraska at Lincoln. Her research interests are nanostructured materials and, more recently, science education, evaluation, and outreach. She has been involved with NSF Experimental Program to Stimulate Competitive Research (EPSCoR) programs, K-12 science education, the APS Forum on Education, and the American Association of Physics Teachers.

Bruce H. Margon, University of California at Santa Cruz

Dr. Margon is vice-chancellor for research at the University of California at Santa Cruz. He was formerly associate director for science of the Space Telescope Science Institute. His research interests are in high-energy astrophysics. As an outsider to NSF and materials research, Dr. Margon brings the perspective of someone involved with NASA science centers and their outreach programs.

Andrew Millis, Columbia University

Dr. Millis is a professor of theoretical condensed-matter physics at Columbia University. His research interests include strongly correlated electron systems, quantum many-body systems, and the behavior of novel materials. He received his Ph.D. from MIT in 1986 and has also worked at Bell Laboratories. He is a fellow of the APS and has been a Fulbright Scholar.

Venkatesh Narayanamurti, Harvard University (NAE)

Dr. Narayanamurti is dean of engineering and applied science and professor of physics at Harvard University. His research interests have focused on electronic materials and the physics of carrier transport in metal-semiconductor devices. Dr. Narayanamurti chaired the most recent decadal survey of condensed-matter and materials physics, and as dean at Harvard, he possesses a broad understanding of the materials research enterprise. He is a member of the National Academy of Engineering.

Ralph G. Nuzzo, University of Illinois at Urbana-Champaign

Dr. Nuzzo is professor of materials science and engineering and director of the Frederick Seitz Materials Research Laboratory (MRL) at the University of Illinois at Urbana-Champaign. His research expertise is in the area of polymers and organic materials as well as chemical processes at surfaces and interfaces of materials. As director of the MRL, Dr. Nuzzo also has broad knowledge of materials research centers outside (although formerly of) the NSF paradigm. He received his Ph.D. from MIT in organic chemistry in 1980; he has received the American Chemical Society's Arthur Adamson Award for Distinguished Service in the Advancement of Surface Chemistry.

Douglas D. Osheroff, Stanford University (NAS)

Dr. Osheroff, the G. Jackson and C.J. Wood Professor of Physics at Stanford University, won a Nobel Prize in physics in 1996. Dr. Osheroff served as a researcher at AT&T Bell Laboratories for 15 years before devoting his time to teaching at Stanford. He is a fellow of the American Physical Society and the American Academy of Arts and Sciences and was elected to membership in the National Academy of Sciences. In addition to the Nobel Prize, he has won many awards, including the Simon Memorial Prize, the Oliver E. Buckley Prize, and the Walter J. Gores award for teaching.

Stuart Parkin, IBM Almaden Research Center

Dr. Parkin is an experimental physicist at IBM's Almaden Research Center in San Jose, California. His discoveries into the behavior of thin-film magnetic structures were critical in enabling recent increases in the data density and capacity of computer hard-disk drives. He is an IBM fellow and manager of the magnetoelectronics unit. His research centers on magnetic materials, magnetoresistance, and thin-film structures. He has received the Outstanding Young Investigator Award of the Materials Research Society and the American Institute of Physics Prize for Industrial Application of Physics.

Julia M. Phillips, Sandia National Laboratories (NAE)

Dr. Phillips is director of the Physical and Chemical and Nano Sciences Center and the Center for Integrated Nanotechnology at Sandia National Laboratories. She is a materials physicist with broad research experience in thin-film growth and interfaces. She was previously manager of the thin-film research group at Bell Laboratories, Murray Hill, and program manager in the Consortium for Superconducting Electronics involving AT&T, IBM, and MIT. She is a past president of the Materials Research Society. Dr. Phillips has been a member of the Board on Physics and Astronomy, the National Materials Advisory Board, and the Solid State Sciences Committee.

Lyle H. Schwartz, Air Force Office of Scientific Research (retired) (NAE)

Dr. Schwartz, retired director of the Air Force Office of Scientific Research, guided the management of the basic research investment for the U.S. Air Force. As former director of the Materials Science and Engineering Laboratory at the National Institute of Standards and Technology (NIST), he managed the 350+ person materials research laboratory including oversight of the NIST nuclear research reactor. He was responsible for the development of the Presidential Initiative on Advanced Materials and Processing. His academic career spanned 20 years at Northwestern University, where he directed the NSF-funded MRL. Dr. Schwartz has received many awards, including the Presidential Meritorious Executive Rank Award and the Department of Commerce Gold Medal. He has been elected to membership in the National Academy of Engineering, was president of the Federation of Materials Societies, is an honorary member of ASM International, and is chair of the board of trustees of the ASM Materials Education Foundation.

Eli Yablonovitch, University of California at Los Angeles (NAE, NAS)

Dr. Yablonovitch is a professor of optoelectronics in the Electrical Engineering Department at the University of California at Los Angeles. He is an expert in optoelectronics, photonic band-gap research and crystals, and quantum computing and communication. He has been awarded the Adolf Lomb Medal, the W. Streifer Scientific Achievement Award, the R.W. Wood Prize, and the Julius Springer Prize. Dr. Yablonovitch received his Ph.D. from Harvard University in 1972. He has most recently served on the BPA's Committee on Atomic, Molecular, and Optical Sciences. He was elected to membership in both the National Academy of Sciences and the National Academy of Engineering.

Neil E. Paton, LiquidMetal Technologies, Consultant (NAE)

Dr. Paton is chief technology officer of LiquidMetal Technologies, Lake Forest, California. Dr. Paton was formerly vice president, technology, for Howmet Corporation, and president, Howmet Research Corporation. He spent 20 years with Rockwell International and 11 years at Howmet. He holds B.S. and M.S. degrees in mechanical engineering from the University of Auckland, New Zealand, and a Ph.D. in materials science from MIT. Dr. Paton was awarded a Titanium Metal Corporation of American Fellowship (1965 to 1968) and the Rockwell International Engineer of the Year Award (1976). He was elected a fellow of ASM International in November 1992. Among recent special assignments, he has served on several National Research Council review committees and was chair of the 1983 Gordon Conference on Physical Metallurgy. Dr. Paton was elected to the National Academy of Engineering in 2002.

NRC STAFF

Donald C. Shapero, Board on Physics and Astronomy

Dr. Shapero received a B.S. degree from the Massachusetts Institute of Technology (MIT) in 1964 and a Ph.D. from MIT in 1970. His thesis addressed the asymptotic behavior of relativistic quantum field theories. After receiving the Ph.D., he became a Thomas J. Watson Postdoctoral Fellow at IBM. He subsequently became an assistant professor at American University, later moving to Catholic University, and then joining the staff of the National Research Council in 1975. Dr. Shapero took a leave of absence from the NRC in 1978 to serve as the first executive director of the Energy Research Advisory Board at the Department of Energy. He returned to the NRC in 1979 to serve as special assistant to the president of the National Academy of Sciences. In 1982, he started the NRC's Board on Physics and Astronomy

(BPA). As BPA director, he has played a key role in many NRC studies, including the two most recent surveys of physics and the two most recent surveys of astronomy and astrophysics. He is a member of the American Physical Society, the American Astronomical Society, the American Association for the Advancement of Science, and the International Astronomical Union. He has published research articles in refereed journals in high-energy physics, condensed-matter physics, and environmental science.

Timothy I. Meyer, Board on Physics and Astronomy

Dr. Meyer is a senior program officer at the NRC's Board on Physics and Astronomy. He received a Notable Achievement Award from the NRC's Division on Engineering and Physical Sciences in 2003 and a Distinguished Service Award from the National Academies in 2004. Dr. Meyer joined the NRC staff in 2002 after earning his Ph.D. in experimental particle physics from Stanford University. His doctoral thesis concerned the time evolution of the B meson in the BaBar experiment at the Stanford Linear Accelerator Center. His work also focused on radiation monitoring and protection of silicon-based particle detectors. During his time at Stanford, Dr. Meyer received both the Paul Kirkpatrick and the Centennial Teaching awards for his work as an instructor of undergraduates. He is a member of the American Physical Society, the American Association for the Advancement of Science, the Materials Research Society, and Phi Beta Kappa; he serves as secretary for the Coalition for Plasma Science and is a member-at-large of the Illinois Mathematics and Science Academy Alumni Association Council.

David B. Lang, Board on Physics and Astronomy

Mr. Lang is a research associate at the NRC's Board on Physics and Astronomy. He received a B.S. in astronomy and astrophysics from the University of Michigan in 2002. His senior thesis concerned surveying very young galaxies in a field beside the irregular galaxy Sextans-A using the Hubble Space Telescope. His mentors were Robbie Dohm-Palmer, University of Minnesota, and Mario Mateo, University of Michigan. Mr. Lang came to the BPA after having worked in an intellectual property law firm in Arlington, Virginia, for 2 years and began at the BPA as a research assistant. He performs supporting research for studies ranging from radio astronomy to materials science and recently received the "Rookie" award of the NRC's Division on Engineering and Physical Sciences. He is a member of the American Astronomical Society.